MANAGING THE HUMAN ANIMAL

MANAGING THE HUMAN ANIMAL

NIGEL NICHOLSON

TEXERE

LONDON · NEW YORK

FOR MY WONDERFUL CHILDREN,
ALICE, NELL, SAM, AND LEO,
AND FOR MY DEAR COUSIN MARION,
AN INSPIRING AFFIRMATION OF
THE LIFE FORCE.

Copyright © 2000 Nigel Nicholson

First published in the United States in 2000 by Crown Publishers, New York, NY

First published in Great Britain in 2000 by

TEXERE Publishing Limited
71–77 Leadenhall Street
London EC3A 3DE
Tel: +44 (0)20 7204 3644
Fax: +44 (0)20 7208 6701
www.etexere.co.uk

A subsidiary of

TEXERE LLC
55 East 52nd Street
New York, NY 10055
Tel: +1 (212) 317 5106
Fax: +1 (212) 317 5178
www.etexere.com

A CIP catalogue record for this book is available from the British Library

ISBN 1-58799-031-8

Printed and bound in Great Britain by Clays Limited, St Ives plc

CONTENTS

Acknowledgments

No book stands alone. Many people have influenced the shape and content of this project, and many more supported me through it. First, heartfelt thanks are due to Mary Nicholson, David Cannon, and Helena Cronin, all of whom read the manuscript at various stages and gave me wonderfully helpful feedback and encouragement. My editor, John Mahaney of Crown Business, also was an assiduous, intelligent, and constructive reader. I thank him and my agent, Helen Rees, for their commitment to the book's concept from its earliest beginnings.

I want also to thank my colleagues at London Business School, for putting up with my bombardment of thoughts from this work and for their comments that often helped me to refine the ideas and their expression. Many other influences should be acknowledged, not least from the inspiration of people encountered at Helena Cronin's Darwin@LSE seminars over recent years. Many of these people are recognized in the chapter notes. Other friends who have consistently supported what I have sought to do here are not so mentioned. I thank them all, for their warmth, belief and encouragement, from start to finish, especially my dearest Adèle.

I would like to make my final acknowledgment to the people of Southern Africa, and their beautiful land, who have been an important backdrop to many of the key intellectual developments underlying this book, from first conception to final delivery. Nowhere have I been more inspired by the warmth of humanity or seen more clearly the human place in the web of life.

MANAGING THE
HUMAN ANIMAL

INTRODUCTION

◆

Instinct in Action

THIS BOOK IS ABOUT the art of managing Stone Age minds in the information age.

We enter a new millennium bedazzled by technological marvels, agog at the possibilities unfolding before us. Voices all around us tell us that there is no limit to what we can do and who we can be. Business media, managers, and consultants take up the theme: by force of will and the exercise of our great intelligence and ingenuity, we can reengineer organizations and management however we wish. We have the capacity to run businesses and make decisions with consistent rationality. We can alter people's motives and cultivate the leaders we need. Politics and turf wars can be eliminated. Rumor and gossip can be banished. Status and sex differences can count for naught.

It is time to get back to reality. We may have taken ourselves out of the Stone Age, but we haven't taken the Stone Age out of ourselves.

By introducing the radical new ideas of evolutionary psychology (EP) to business, I want to help put an end to utopian dreaming in management. We can get a lot more done by working with the tidal flows of human nature—understanding what motivates people and the limits to change—than we can by managing as if we can to stop the tides from coming in. Managers get dispirited, exhausted, and cynical from trying to do the impossible.

Evolutionary psychology shows a new way forward, through

a science-based approach, built upon a great and growing convergence of evidence from many disciplines. It reminds us that we are animals, with a highly engineered, genetically encoded design, a design as much for our minds as for our bodies. Mind and body as a single organic identity were shaped by millions of years of evolution for a particular kind of vanished existence. For almost our entire history this was the life of seminomadic clans, hunter-gatherers of the savanna. We stopped evolving long before we started building cities, organizations, and workplaces. And now we face the consequences of misfit between our ancient nature and the society we have created.

Our brains, hardwired to help us survive and reproduce in an ancient world, do things such as:

◇ Make snap judgments based on emotions
◇ Let one piece of bad news drive out a hundred pieces of good news
◇ Take big risks when threatened and avoid risks when comfortable
◇ Allow confidence to conquer realism to get what we want
◇ Create opportunities for display and competitive contest
◇ Classify things and people, dividing groups into "us" and "them"
◇ Practice gossip and mind reading as key survival tools

The message of this book is not that we have to passively accept the costs of our instincts in the modern world. We need to understand them so we can manage them with insight.

This book is about business, but remember it is not a separate world. In human life there is only one realm of experience. We love, hate, work, and play everywhere. They might be expressed differently, but the same impulses move us in the home or in the office. So here's a story about a real event that happened to me on the journey from work to home. Looking at this story through the lens of evolutionary psychology, we can extract themes about the nature of human nature and see how these relate to what we see every day in our offices and factories.

It was six-thirty at night in the late fall: the sun was just setting. I had just left the office, picked up a few items at the corner drugstore, and was walking toward the corner of the street where I lived. I was walking in the shadow of a boarded-up building, a small dead zone in an otherwise pleasant residential area. Coming toward me was a group of five young men in their late teens or early twenties. I thought nothing of it as I strolled along—briefcase in one hand, groceries in the other—until I saw the men fan out as they approached me. At this, a light of recognition ignited in my brain that this situation was unpredictable and dangerous. Strangely enough, I had no sense of fear. This new scenario seemed at once curiously familiar yet unreal, as if I were an actor in a drama. What I had to do was clear to me. I had to keep walking straight ahead as if nothing was happening.

In a minute we were passing each other and I was in the midst of them, two in front, one on either side, and the fifth out of view behind me. One of them said something to me. It sounded like a question, but I didn't hear the words. I knew what was happening, but my course was set: keep walking. Still quite calm. Suddenly I felt my head gripped in a hard armlock from behind and found myself bumped down roughly onto the sidewalk, my bags scattered around me. This was a surprise. Yet still I knew the rules of this game. "Take my wallet," I said through strangled throat. "Where is it?" a voice asked. I pointed to my trouser pocket. My wallet was swiftly removed and they were off running away from me. I hauled to my feet and realized they had taken my briefcase as well. This worried me. It contained my Filofax, the most vital element of my baggage. I stood and yelled at them in what I felt was a strong and reasonable voice: "Please leave my briefcase! There is nothing in it of any value to you!" I called this out twice. I felt they owed me. After all, I had played by the rules of the game and given them my wallet. They should give me my briefcase! At the end of the street, just before they disappeared—apparently in response to my second call—they paused and then dropped my precious briefcase.

I calmly walked up the street, retrieved it, and then walked home, where I immediately phoned first my credit card companies and then the police. Only then did I notice that my wristwatch was also missing. They had removed it from my arm, and I hadn't even noticed. Some twenty minutes later the police arrived, and I recounted my story to them. Only at this point did I experience unexpected and powerful emotions. I found myself shaking. The police were very reassuring. "You did exactly the right thing," they said. "If you had resisted you could have found yourself knifed or worse. We know these gangs."

I reflected. Yes, of course they were right. But it didn't feel right. I felt as if I had done something wrong. I should have known the risks. I should have foreseen the situation and avoided it. But how could I have? I could see that these thoughts were not rational. How could I possibly have anticipated what would happen? It was all over so quickly. But the thought persisted: I had done something wrong.

This is a thoroughly modern urban story. While we could analyze it in terms of the lifestyle and motives of deprived youth in the inner city, we have bigger fish to fry. In this story we can find deeper and more timeless themes about who we are, why we act and think the way we do.

Emotions Before Reason

It might look as if my reason controlled my emotions throughout the episode. On the contrary, my emotions controlled my reason. At critical points I did not stop and think; I just acted instinctively. I did what felt right. Emotions, a constant magnetic force in the mind on all our thoughts and judgments, have their own wisdom. Throughout the incident my bloodstream was awash with powerful emotion-controlling chemicals working in conjunction with a rush of perceptions and unconscious calculations about how to act. The reason of the heart, not the head, suppressed all conscious emotion until it was safe to do so. I didn't even notice my watch being removed from my wrist.

See what happens in the office or factory. People are acting unemotionally even when under provocation that would justify an outburst. Instead people act impassive until it is safe to discharge feelings even though their emotions are actually influencing their behavior continually. When feelings do erupt, it is easy to lose control. The annual Christmas party has found many a boss shell-shocked by the emotional release of the unhappy staff, whose feelings, with the aid of alcohol and atmosphere, have slipped the leash. Even when we think our reason is in command, it is dancing to the rhythms of the emotions.

Confidence Before Realism

As the event unfolded I was unafraid—irrationally, for this was certainly a situation to be fearful in. Instead I walked into it with an extraordinary sense of invulnerability. We are hardwired to tough it out in situations where we cannot avoid challenge. In fact, we seek out challenging situations, like competitive sports, with the very aim of arousing the psychic buzz of confident performance.

In business this has big risks. Many a chief executive has suffered through this hubris—the arrogance of leaders who think they are gods, just before they fall. Overextended projects, improbable mergers, premature product launches, bad appointments—these and many other follies derive from nothing more than executive ego. So strong is the confidence trick that when we feel good we actually ignore or discount the cues that might tell us we are walking into a disaster.

Loss Aversion and the Quest
for Control

At first I was not aware of either loss or threat. The first sensation of loss came when my attackers ran off with my briefcase. It was only then that I got excited. Had they been close, I might have acted impetuously in the attempt to retrieve my precious case. But the real strength of loss aversion reaction showed in my psychological reaction afterward. I experience this as a powerful desire to reassert control, here by thinking about what I had done wrong and how I could avoid its happening again. When an experience is traumatic or painful, we strive to reestablish a sense of mastery. This drive underlay my irrational feeling that I had made a mistake and my magical thinking that I could have avoided the unavoidable, foreseen the unforeseeable, and controlled the uncontrollable. This is a kind of psyching up after the event. In the world of our ancestors, where life was tenuous and advantage marginal, to lose something of value could be to lose it forever. Evolution has sensibly programmed in our brains a much stronger drive to avoid loss than to seek gains. This reveals itself in many ways, one being our vulnerability to magical beliefs about how we can avoid chaos and danger.

In business, our desire to avoid loss surfaces as failure aversion. Research shows how obsessed people in organizations are with failure and its avoidance. We can see the consequences all around us in the cover-ups, delusional thinking, and emotional breakdowns that accompany disasters where human action is partly to blame. Look at most corporate disasters—for example, the collapse of Barings, one of the United Kingdom's oldest merchant banks, through the errors and misdemeanors of one of its young traders—and you will see that a major driver of how people mismanage such events is their aversion to loss (usually trying to save face).

Classification Before Calculus

It's difficult to know for sure what was going on in the heads of my human predators, but we can make some informed guesses. It is very likely that they had no thought for or interest in what it might be like for me. I was an impersonal quarry. They were not calculating the odds against them, either, such as whether they would be apprehended or how I might react. They were a group, and I was an opportunity. To boys from the other side of the tracks, I was instantly classifiable as middle-aged professional astray from his herd—fair game! When we make decisions, our brains are programmed to make quick and efficient classifications of people and situations, which can then trigger the action that is most fitting. People often compare the brain to a computer, but it is lousy at computing. We seldom calculate probabilities, and when we do we are pretty hopeless at it. But classification is second nature to us. If you want to give a good lecture that people will remember, include lots of memorable lists; don't clog them up with abstract logic and calculus.

In business this bias is visible constantly. If you want a group to come to a particular decision, just try to manipulate things so that option A (the one you favor) is classified as good, wholesome, or safe, and that options B to whatever are seen as eccentric, difficult, or dangerous. In a recruitment round, to get your favorite candidate selected, get all the candidates you do not favor tagged with negative labels. Then construct a story about why your candidate is right. This builds on an associated principle of memory and meaning: *narrative before accuracy*. Our brains are constructed in such a way that accurate recall is almost impossible except for coded lists and facts. Anything dynamic gets changed in the mind from the first replay to the latest; at each telling the story gets a little more refined so as to make more sense to us. And anything concerning people gets recorded and replayed as a dramatic reconstruction having to do with motives and feelings. This is how we are designed to process information: to give point and purpose to our actions, not to be impassive and accurate video recorders.

Contest and Display

I didn't have much time to check out how my assailants were interacting with one another, but eyewitness accounts of gang life tell us that events such as these serve familiar purposes for the whole group and its individual members. Little dramas are enacted within groups by people performing familiar but unscripted roles in ways that both unite them and distinguish them one from another. Their participation is unequal. Some individuals show their leadership by taking an active or dominant role. This role may be contested. The initiation of my mugging might easily have been a bid by individual X for regard and status in competition with person Y. These people are much more interested in each other than they are in me. I am just the raw material for their group work. Being able to compete with one another in a situation where they can all feel like winners (after all, they got away with my wallet) enables them to feel simultaneously individuated and integral to the group, even while jockeying for position or regard.

Sports and entertainment institutionalize the human desire for public contests and display of this kind. The same drive also mobilizes many behaviors at work. We even create arenas for its unfettered exercise. Career systems are full of contests and rituals, such as the stylized way interviews are conducted. In marketing, "beauty parades" and ranking tournaments occupy a huge proportion of people's energies. In professional life, conferences are set up like Roman amphitheaters for public shows and dramas.

Group Discrimination

One type of classification has especially potent and pernicious effects: seeing people as in or out of our own group, often in relation to how visibly different they are from us. This forms the basis of all the discriminatory isms around race, sex, age, or any other distinguishing characteristic. It takes very little to awaken an innate programming that is designed to support kin and near kin against strangers. At its extreme, as in my story, you get gangs operating like hunting parties.

Look around your organization and you will quickly find groups who define themselves and their interests as superior to others. Each one seems to need another to look down upon. The powerful (finance) are demonized. The weak (human resources) are satirized. Close neighbors deride one another—sales vs. marketing, production vs. design, front office vs. back office.

Gossip, Narrative, and Grooming

You can be sure that the gang retold the tale of the mugging a few times to one another and to others outside the group, all with suitable embellishments to make it seem less mundane than it was. Maybe, without intending to, I have done the same in telling the story to you. The whole point of many an initiative is simply to give us something to talk about. Even when we are cast in someone else's drama, we still look to extract some social dividend from recounting our own version, for we love to tell and hear stories. For this reason you will get plenty in this book. They help us to understand our lives. They contribute to our ability to feel like competent actors in life's dramas. Stories serve other purposes, too. They fill gaps. Rumor is improvised news. Gossip is grooming, a service people render to one another. We tell each other tales to signal inclusion, to show we like and trust the person we are telling. Gossip is also a mainstay of everyday politics: in the game of reputation we gossip about each other to influence how we and others are seen.

Offices are full of gossip and rumor. People trying to run a business along lines of controlled accuracy may find this infuriating, but they would be missing the point that workplaces are communities, and people at work continue to try to behave as if they were villagers rather than factory hands or bureaucratic functionaries. Thank goodness. The color and life of work experience come from the our insistence on not behaving like robots.

Hardwired Individual Differences

Of course, the way I reacted was personal. It said something about the way my brain is wired that I didn't get mad, make a run for it, burst into tears, or do something crazy. Other people might have reacted differently. Even if they didn't and they acted just the way I did, they would still have quite different sensations, feelings, and afterthoughts. The same is true of my assailants. Within a gang, the experience of each individual is different from that of any other. Individuals act out their subtly different roles under the direction of their unique character biases. Some of this individuality comes from what we have learned over our lifetimes, but increasingly science is discovering how much of the range of human reactions is genetically encoded and written permanently into our hardwiring.

At work every day we can see people reacting to events and behaving in ways that we wouldn't dream of emulating. Our individuality has important effects. Leadership is a prime area. Big shifts in business strategies often result from the personality of a new boss.

And one hardwired individual difference affected all of us together in my mugging scenario. My assailants and I were male. Several aspects of the drama we acted out would have been almost unthinkable for females. Many of the problems we face in business are because of hardwired differences between male and female psychology and the fact that most businesses are run to satisfy distinctively masculine drives.

The Trading Instinct

At the end of the scenario I did a trade with my assailants. I had played their game. It made me feel that I had a right to demand that they play fair with me and return my precious briefcase. The fact that they did so suggests that they too understood that we were doing a deal. Afterward, the situation seemed familiar to me, a half-forgotten childhood memory, and then I remembered: it reminded me of the gang behavior of boys in the schoolyard and the kinds of implicit rules that everyone there seemed to understand and stick to. This is what "honor among thieves" means. Evolutionary psychology says that we are deeply attuned to implicit contracts and expect them not to be violated, even when things look chaotic on the surface.

Executives verbally brawling at a conference generally stick within quite tight boundaries of conduct and often end up doing deals. Our reactions to violations of implicit contracts are often very fierce. When employees feel their company has reneged on the spirit of an agreement, they feel a bitterness often out of all proportion to any material loss.

The "As If" Principle

The episode also illustrates the "as if" principle. All relationships have primal origins. We first learn about other people from growing up within a family. The models of parent and sibling relationships are reproduced in other contexts with nonkin. For example, when we treat our boss as if he or she were our father or mother and respond to our colleagues as if they were our siblings, the "as if" principle is being triggered. In fact, we assess almost all social encounters from a framework of basic ways of dealing with kin and nonkin. My mugging encounter involved nonkin on all sides, but they, the gang, acted as if they were a tribal clique encountering a stranger. I, for my part, acted out the role of the outnumbered newcomer or stranger to the clan—surrendering without resistance, being polite.

In management, one can play on primal models of people's experience to produce the best results. But often we do the opposite. Bosses treat subordinates as incompetent and dependent children instead of as adults needing trust. Peers treat one another as sibling rivals instead of figuring out how they can advance their common interests as buddies.

As these themes suggest, evolutionary psychology can supply us with new lenses to view the world. By applying this vision to business and management, we can view afresh what motivates managers, leaders, and employees. We can see more clearly which methods of management and organization go against the grain of human nature. We can find new ways of working.

Evolutionary psychology (EP) shows us that what seem to be very local contemporary social events have been experienced by earlier generations since time immemorial. Because these experiences are cloaked in the manners and appearances of our times, we fail to see them as what they are: dramas with an ancient lineage.

This is the starting point for the new discipline of evolutionary psychology. Its mission is threefold.

First, EP seeks to understand the evolutionary pressures that created us and what those pressures did to shape the structure of the human mind.

Second, EP uses this knowledge to explain how our evolved nature guides our actions and colors our experience of everyday life.

Third, EP considers what happens when our minds encounter new circumstances for which they were not designed—that is to say, how we adapt and try to act out our nature within the modern world.

In this book I want to turn this analytical spotlight on what is happening in our workplaces. It can help us understand what makes the difference between the most impressive examples of leadership, organization, and achievement and the most ineffective, unhealthy, and destructive. It can help us appreciate the limits to change. But it is important to note that EP's insights do not eliminate choice. On the contrary, they help us to focus our decisions where they can have best effect. A true understanding of human nature—what motivates people and shapes their thoughts and actions—can enable us to build organizations fit for humans and instruct us in how to manage one another to bring out the best in everyone.

Chapter 1 sets out the basic ideas of evolutionary psychology and the scientific evidence underpinning it. This is the story of how our sophisticated design came about, what its principal features are, and why they matter.

Chapter 2 illustrates how these same features can create the dark side of organizational life. We will look at the seven deadly syndromes, some of the worst features that keep recurring in contemporary business. We will consider what makes them happen and how they can be avoided.

In Chapter 3 we look at one of the fundamental themes of evolutionary psychology—perhaps the most controversial. This is the differences between the sexes and the problems we create by imagining that they or their effects can eliminated. Acknowledging this reality means we can look for new ways of transcending the male organizational paradigm to get the best from both men and women at work.

Chapter 4 revisits one of the most written about and least

well understood themes of business: leadership. Evolutionary science gives us a fresh take on this in several ways. It shows the deep origins of our obsessive interest in and the mythic qualities attributed to leadership. It also shows how inborn differences in temperament make people either fit or unfit for leadership positions and how organizations tend to get the leaders they deserve because of their structures and cultures. This leads us toward new ideas about what we can and can't do to improve the supply of good leaders for business.

In Chapter 5 we look at the human mental tool kit in greater detail. We explore exactly how our ways of thinking and feeling are programmed for a particular kind of existence. In the modern world this programming can lead us into some disastrous errors. Insight into our hardwired biases can help us avoid these.

Chapter 6 moves into the realm of relationships. The models we call upon are primitive in several senses. What I shall call "gene politics" are the ways in which kinship governs interactions within human groups. Dealings with people outside the family also follow ancient protocols of alliance and conflict. Daily dramas are played out in family firms and work groups that reflect our prototypical relationships. Understanding these dynamics can lead us toward more harmonious ways of relating.

Communication arts are prime human skills, and in Chapter 7 we look at how they function in working communities. We look at the contrast between approved channels and the well-worn paths of gossip and informal networks. Some managers mistakenly think they can control the flow of information at will and see gossip and rumor as destructive forces. But in fact, informal communication is the lifeblood of all communities, and we consider how businesses can turn this to their advantage.

Chapter 8 puts all these elements together to look at how organizations function as communities. They come in countless different shapes and sizes, but running through all of them are common themes. We shall look at two such themes: how people react to different kinds of business structures and what forces underlie corporate culture. Leaders have a special responsibility for how these reveal themselves.

My concluding thoughts are on the future of business and management in the twenty-first century. Far from moving us back to traditional forms of business, EP can point the way toward new and better strategies. Aided by the tools of the information revolution, evolutionary psychology can draw the blueprints and make a reality of new ways of managing and organizing human communities.

◆ I ◆

Stone Age Minds in the Information Age

UNNATURAL MANAGEMENT

Ask MANAGERS what they find most difficult and challenging in their job, and you will get one of two replies. One of the most common (if they're being honest) is a one-word answer, usually after an embarrassed pause: "people." Another frequent response is "the management of change."

Now, isn't this curious? Why should we find most difficult the very arts one might say we were created for? We are the most sophisticated of creatures when it comes to social intercourse. As for coping with change, our powers of adaptation are prodigious. We have supreme skill as learners. Dealing with people and change should be a cinch. The answers we should expect people to give to the question might be such tricky matters as doing accounts, writing reports, or any number of other things that it takes us painstaking hours of practice to learn, and even then we often do badly.

So what's the problem? The problem is this: organization makes us do the things that come naturally to us in an unnatural way. This creates a tension. We make a mess of managing people and change because we try to do them according to the rulebook, following the constraints of our business worlds. This turns out to

17

be hard going, and then we get into difficulties because we keep backsliding into instinctive ways of acting. The result: stress and confusion.

Take an example: I have to tell someone that he or she is not performing. If this person were a member of my family, maybe I could do it in a loving way. If he were my good buddy, maybe I could speak with him in a way that would help him figure out for himself what's wrong and what he could do about it. But here before me stands a comparative stranger (not a friend or a relative) whose annual pay and future career depend on what I say about him. No wonder we both find it embarrassing and I handle it badly. It takes a lot of practice to do well the unnatural things that managers have to do every day.

Let's take another example. I'm walking through the office, and a colleague's secretary stops me and asks for a confidential word. OK, I say. She tells me about a big problem she's been having with her boss, who is a friend of mine. Would I help? Foolishly I agree (I should advise her how to sort it out for herself). I go to my friend and gently suggest that he might try working with her in a different way. He erupts and tells me to mind my own damn business. I got my fingers burned by making the mistake of acting as if I were part of a community rather than sticking to conduct appropriate for a node in the hierarchy: sticking to the approved channels.

I could also recount numerous instances where change management has failed through such dislocation. Especially common today are breakdowns of new technical systems. Managers buy complex systems in hopes of improving efficiency and productivity. When the systems go wrong, managers in their ignorance are quick to blame the technicians, who blame weaknesses in the system or lack of resources. Soon people start protecting themselves and resisting new innovations. The managers then wash their hands of "technical" matters.

Change involves the politics of the community. In a world in which people have staked out territorial claims for themselves around roles, resources, and psychological investments, why would anyone want to cooperate with changes that reduce their status and influence? The internal structures of businesses are formed by people molding jobs and systems as much as possible to suit themselves—and then of course clinging on when anyone tries to change the architecture. Even when customers will benefit and money will be saved, turf wars prevent change from happening.

There is another way of managing change. Small businesses and some units within large firms, such as project groups, embrace change as exciting, challenging, and beneficial. They do this by following the fluid patterns of self-organizing teams within a community, such as a tribal hunting party or entertainment troupe. Individual contributions are valued. Experimentation is encouraged. Errors are analyzed, not blamed. People give to one another. Everyone shares a vision. The group's work is valued by the community.

In the business world you can find numerous examples of creative organizations, constructive human relations, inspired leadership, and dynamic change. Within even the most deadening bureaucracies you can find pockets of vibrant innovation and communal spirit. Sadly, it is often much easier to find the opposite: inefficiency, unhappiness, and frustration. For many managers and employees, business as usual is a daily catalogue of irritations and upsets, relieved by friendships, leavened by humor, and, from time to time, compensated by achievements.

What's the difference? On the whole, when things go well, we are free to operate in tune with our instincts, and when they don't, we are not. It's not quite as simple as that. Often the bad stuff happens when we are trying to act natural in unnatural circumstances. If we could be true bureaucrats we could make bureaucracy work, but the natural politician in us keeps barging in and mucking things up.

Problems also come from starry-eyed idealism. When we try to create utopias in which people are supposed to be uniformly creative, free from envy, blind to status or sex differences, and unbiased in their judgments, we find human nature reasserting itself, tugging us back to our normal states of greediness, pride, self-centeredness, and sloth.

The most respected managers, leaders, and organizations have figured out, usually unconsciously, how to put human nature to best use. You will notice that the people we rate as the best natural managers are those who seem to understand, tolerate, and even enjoy the foibles of their staff. They are clear and pragmatic in how they act. They seem to understand themselves and their limitations. They don't expect situations to be perfect, but they progress toward solutions without fuss or show. They always work with rather than through people. They give and receive trust naturally.

The secret is simple. It is the recognition of what we are: what we have in common and how we differ from one another. So let's start by looking at what the new science of evolutionary psychology is telling us about what it means to be human.

THE ANCESTRAL ENVIRONMENT

EVOLUTIONARY SCIENTISTS have established the essentials of human history. Our line started around 4 million years ago, when global cooling created a new kind of savanna environment. This was the ecological niche into which our oldest direct ancestors, the first two-legged clan-dwelling apes of eastern Africa, emerged. Over time, evolution progressively refined the prototype and greatly enlarged its brain. The last stop on this line was modern man, *Homo sapiens,* who emerged around 200,000 years ago, fully adapted for our ancestral environment, the world of clans of hunter-gatherers on the savanna. By 45,000 years ago, modern humans had spread all over the planet. Our lives have changed immeasurably since that time, but our genetic constitution has not. In fact, the human genetic profile is very tightly specified: there is less difference genetically between, say, an Australian Aborigine and a European aristocrat than between any two chimpanzees. Our minds and bodies are those of Stone Age people. Superficial local adaptations such as skin color have taken place regionally, but our core mental functioning is singular and constant. What we were then, we are now.

What we were then was seminomadic clans of up to 150 members, bound together by ties of blood and a sense of shared fate. The wide geographical dispersion of these communities would have made encounters with strangers rare events. The known world to which individuals had to adapt was a community of working families, interwoven by kinship and regulated by informal contracts. Status maintenance and enhancement would have been ever-present concerns for everyone. Leadership was an accepted fact of life, but leadership roles for males would have been multiple according to the demands of clan life: young chieftains leading hunts; elders resolving disputes between clan members. Leaders would have enjoyed privileges and resources, including the best opportunities for mating and reproduction. Their positions would have been reasonably secure in the short term, but natural hazards and challenges by new contenders for power would have made leadership provisional and fragile. High-status women would also have exercised power by influencing critical decisions and determining key alliances.

The organization of these groups would have been democratic and fluid, according to who was available, possessed the relevant skills, and could be trusted. Around the necessary tasks—long-distance hunting, local foraging, toolmaking, preparing food, caring for the children, training the young, administering justice, creating art, conducting religious rituals, and entertainmenting one another—existed a flexible division of labor. However, many of these tasks would have been specific to one sex or the other. Men and women would have had distinct responsibilities—for example, men almost exclusively in long-range hunting, women in child care. In other tasks, such as toolmaking and foraging, men and women would intermittently have worked together cooperatively.

EVOLUTIONARY PSYCHOLOGY: THE SCIENTIFIC EVIDENCE

SINCE WE have no way of observing directly what life was like among our Stone Age ancestors, we have to piece together a picture from the scientific evidence, a picture that is continually being refined as new evidence comes to light. What is remarkable is that everything we are discovering from a wide range of completely different disciplines all points in the same direction.

Darwin's Framework

Charles Darwin's insights about evolution are the bedrock of all subsequent investigation and discovery. During his travels he discovered many new species and reached the conclusion that these had somehow been shaped to fit the environments in which they lived. Natural selection is the simple but powerful insight that creatures possessing a feature that helps them survive and reproduce go on to multiply and displace those that lack such an advantage. In this way, over successive generations species' abilities come to fit their way of life: for example, nocturnal hunters possess acute hearing and specialized night vision; predators develop speed, power, and weaponry. Some creatures become highly refined specialists adapted to a localized niche, such as the giant panda, which can survive only in bamboo groves. Others become generalists, such as scavenging rodents, which can prosper under many conditions. Darwin noted that the principal gift of primates was social intelligence, the ability to gain mastery through organization. This is what made his idea "dangerous" in the orthodoxy of his Victorian times. It implied that we humans were shaped not by divine intervention, but by twisting path of natural selection. We too are animals, and at every turn we can see how human behavior comes from our animal essence.

The New Darwinism

Darwin's genius was his ability to develop a viable theory without any knowledge about the mechanism that makes evolution work: genetic inheritance. We now know that evolution works through the genetic coding that governs the biology of every organism. The "selfish gene" contains the code for each individual's unique DNA profile. Spontaneous mutations, such as copying errors at the point of fertilization, alter the program randomly. If the mutations have a beneficial effect, the individuals containing altered genes will survive to reproduce themselves. If, as is more likely, the mutations are harmful, they extinguish their host.

But new genes have to do more than help us compete for food; they need to enable us to pass them on, to reproduce. This is the idea of sexual selection, a concept that has been elaborated by biologists since Darwin. It says that the refinement of species' characteristics continues beyond the basic design through sexiness, the ability to secure mating opportunities. If you can produce more offspring or offspring of better quality, whatever features make this possible will be passed on to the next generation. This leads to some strange results, features that would seem to have no survival value. Classic examples are the male peacock's gaudy tail feathers and the bull moose's monstrous antlers. Both impede for feeding and maneuvering, but are irresistible to females. Why? It is now reckoned that there is a method to this madness. If you are strong and healthy enough to carry a big handicap, it is likely you have good breeding potential and therefore will have "fit" offspring. Your good genes will be passed on. When males and females are searching out signs of fitness and thus the best chances of reproducing, handicaps have to be genuine to guarantee against false advertising. To be able to bear their useless burden, good genes are required.

It has been reasoned that this idea explains much human behavior. We spend inordinate amounts of time and energy making ourselves look good in various ways. Painstaking self-decoration, lives devoted to refining skills beyond our immediate survival requirements, ceaseless work, and competition to acquire ever greater wealth—sometimes to the point of being self-defeating—are visible all around us. Many puzzles are solved by this idea: the millionaire entrepreneur who continues to chase money and never has the time to spend and enjoy it; the scientist looking for a wonder drug who works himself into an early grave; the diva who sacrifices her family life for the sake of her career; the bond dealer who buys a 140 mph Ferrari to drive around city streets at 15 mph. All are the product of runaway motives to display.

The Neuropsychology Revolution

At the same time as biologists are decoding the human genetic blueprint, neuroscientists are mapping the mind. Until recently, the conventional wisdom was that humans differed from other animals in possessing minds like blank slates, on which learning and culture could write the story of human nature. It told us that every newborn baby has her psychology inscribed by how she is raised. We now know that this is profoundly wrong. Far from being a general-purpose computer, the brain is a heavily hard-wired library of programs that shape our identity. Research in genetics and neuropsychology, using new technologies such as brain imaging, is revealing the intricate specificity of this design.

Genes for individual abilities, tastes, and motives, as well as a range of psychological disorders, such as alcoholism and depression, are being identified. At the same time, we are learning that particular areas of the brain are wired to govern our experience of emotions, sensations, perceptions, our use of language, and the kinds of control we can exert over states of mind and body.

Almost everything we do and think relies on our biochemistry and nerve centers, laid down for us by our genetic programming and activated on a prescribed schedule of maturation or via exposure to critical experiences.

Scientific Convergence on the History of Mankind

Astronomers and physicists look at the stars and deduce the history of the universe. Geologists analyze rocks and construct the story of the planet. Science has an impressive track record in recreating what is long past, and new techniques build our knowledge ever faster. In the same way, with the aid of increasingly sophisticated tools, scientists from different disciplines are closing in on perhaps the most compelling mystery of all: the origins of the human mind. Paleobiologists study fossils and deduce our place in the development of life on earth. The remains of our ancestors tell the story of great changes that occurred in the human lineage. Archaeologists join the hunt. By examining the remains of prehistoric human settlements, such as tools and primitive art, they are able to reconstruct how we once lived. Climatologists tell us about how shifts in weather patterns shaped the environment to which we had to adapt. Anthropologists study lost civilizations and cultural groups living today and find universal themes in the life of the human tribe. Primatologists find parallels between these themes and what happens in the societies of our closest relatives, the great apes. Biologists uncover the mechanisms governing behaviors we share with a great diversity of other species.

A key conclusion has emerged from all this evidence. This is that the profile of human nature was fully delineated long before the dawn of recorded civilization. Over the past few millennia, no single environmental pressure—such as global climate change—has pushed the evolution of our species in any new direction. The period since we made the transit from hunter-gatherer existence into a life of crop management and stock breeding in fixed settlements—around 10,000 years ago—is too short an interval for anything other than superficial variations (absorption of lactated milk, skin pigmentation) to occur. In fact, compared with other species our genetic profile is very narrow in its bandwidth of variation, due to winnowing of variation from bottlenecks of near extinction in our history. Also, by 10,000 years ago we were scattered across the globe and rather numerous. These are conditions favoring local variation, not a wholesale shift in species evolution. No significant mutation to our mental architecture could have been able to get a grip and spread. Almost all our psychology has an ancient lineage that we shared with our ancestors and that we continue to hold in common with other species, especially other primates.

What has changed for humans is culture. The recent invention of agriculture was the change in our story line that led to all the problems and dilemmas of modern life. It was almost just as Genesis tells us. Out of our hunter-gatherer Garden of Eden we sallied forth into a world of toil. Once you start cultivating crops, you can support a population a hundred times larger on the same territory, and you don't have to go wandering to find your next site for hunting and gathering. The catch is, you have to work harder—much harder. The more you can produce, the larger the population grows. The more people you need to support, the harder you have to work to provide for them. Thus the vicious cycle of modern existence started to turn. To sustain an agricultural society you need discipline, organization, and relentless labor. As God said to Man in the Garden, "In the sweat of thy face shalt thou eat bread."

So why did we let this happen? The answer is that our genes do not allow us to be content with what we have. They push us in search of wealth and improvement. Individuals and groups compete for the best resources, as they always have. But now in an agricultural economy you could lay up your store, unlike in hunting and gathering societies, in which most wealth was perishable. Accumulated wealth gives power to command. Moreover, you can pass your store on to your kin to secure their future and thus the longer-term selfish interests of your genes. Men loved it. So did women. Here was the best chance for your genetic inheritance: get wealth and pass it on to your own.

Thus was modern society born. The steps from then to now have been short and rapid. Large agrarian settlements developed quickly. By around five thousand years ago, the first cities had sprung up. The rest is recorded history. The Sumerians, the Egyptians, the Greeks, the Romans, the Chinese, the Mongols, the Goths . . . the Wall Street bankers—all in an eyeblink of human time.

MYTHS OF MANAGEMENT

WE HAVE achieved so much so quickly. It is hardly surprising that we have become infatuated with our powers. And indeed, we are wonderfully inventive, ingenious, and adaptable creatures. Many as-yet-unimagined wonders will be wrought through our galloping technological inventions. More and more we are finding that what we can envision we can build. But the key question is, can we make our innovations work? If these depend on humans—the thoughts, feelings, and actions of an operator, manager, or consumer—we have to ask if we humans and our evolved nature might not be the limiting factor to achieving our dreams.

We can design all manner of organizations and social systems, but can we make them run? Some people thought communism a great solution to the problems of oppression, greed, and conflict, but communism failed at all its attempts. The Israeli kibbutzim were based on the ideals of communal living and child rearing and democratic, nonhierarchical authority. Kibbutzim still exist, but not in the spirit of their original design. Traditional family models of living inexorably reasserted themselves, and informal hierarchy and division of labor sneaked in through the back door. Throughout history our impulses and instincts have undone our idealistic designs for ways of living and organizing. Communism and the kibbutzim were based upon false views of human nature. (As were fascism and military dictatorship, lest anybody think this argument one-sided!)

But can we not change human nature? Every parent would like to believe we can. Yet even the most tireless efforts to imbue our children with stainless moral character, positive emotions, and intelligent sensibility are undone by the flaming recurrence of our offsprings' animal passions. And when you have more than one child, you can see it right away. Two children from the same parents differ from each other in ways that have nothing to do with what parents and the world have done to them. It's in their genes. So what is left for mom, dad, and teacher to do? Quite a lot, actually. If we have insight into human needs and capabilities, we can build environments that bring out the best in one another. We can give each other powerful and life-enhancing tools—ideas, objects, beliefs. We can teach skills. In lots of ways we can shape behavior, but only within the limits set by the human design.

For many people—especially the young, who want to believe they can do or be anything they like—this is an unwelcome message. In talking about these ideas to groups, I find most resistance among young MBA students. This is not what they want to hear on the brink of their glittering careers. Indeed, young people's belief in infinite possibilities may itself be a hardwired adaptive mind-set. Older people, sadder and wiser executives, quickly get the EP point about the limits to change. It brings a new level of understanding to their years of confrontations with their own impulses and with other recalcitrant humans. But hope is hardwired in us, and even case-hardened managers are prey to the belief that there are magic formulas for aligning people with strategies, technologies, and organizational designs. The myth has a double appeal: profitability as well as perfectibility. It is in our nature to believe in magic.

I have to admit that my own profession of management education must take a share in the blame, along with all the advisers, strategic planners, and consultants who have built an industry on optimism. It is one thing to believe we can reinvent our institutions to fit our goals. It is much less reasonable to think we can do the reverse—reinvent ourselves to fit our institutions.

We are encouraged in this error by observing the striking differences between people from different cultures and classes. From this perception it is easy to draw the false conclusion that our species is infinitely malleable. All it really proves is that human nature can dress up in lots of different ways. Below the surface, recurring patterns are visible everywhere, as people follow their primeval drives and ways of thinking.

Here are some of the common beliefs about management and organization and how they stand up to the view of human nature the new science suggests:

MYTH: Differences between men and women in work are solely the product of cultural brainwashing.

REALITY: Men and women have different hardwired psychologies, so it is normal for them to want to do different things and to do the same things in different ways.

MYTH: Rational decision-making and unemotionality are normal in the work world.

REALITY: Our emotions always have their hand on the mind's tiller, directing the focus and biases in our logic. It takes a lot of effort for most people to stay coolly reasonable, and we can maintain our rationality for only limited stretches of time, and with the aid of tools and disciplines.

MYTH: We can get along without leaders.

REALITY: We can abolish formal leadership roles, but informal leaders spring up naturally on every side. Some people will always want to be leaders, and others will always want to follow them.

MYTH: Anyone, given the right training and experience, can be turned into an effective leader.

REALITY: Individuals' unique genetic programming makes some people more suitable for leadership roles than others. Some are definitely born *not* to lead.

MYTH: We can create nonhierarchical organizations.

REALITY: We can do away with the formal structures of status hierarchy but can never eradicate informal pecking orders of power and influence. People, especially men, will always establish hierarchical relationships by competing for reputation and status.

MYTH: Most people are risk-averse and resist change.

REALITY: People are loss-averse but will actually seek out risk and embrace change, especially if they can see how they stand to gain or how they might avoid future loss.

MYTH: Politics and corruption can be eliminated from business life.

REALITY: People will always be motivated to give and return personal favors and allow friendship to influence decisions.

MYTH: Through cultural conditioning, we can alter employees' core motives and values.

REALITY: We can change people's behavior but not the deep structures of their personality. These grow out of inborn profiles of temperament.

MYTH: Employees can be induced to identify with their corporation.

REALITY: We can get people to feel positive about the corporate brand, but people identify first and foremost with the small groups they belong to. Only through such groups can you usually get people to make sacrifices for the business as a whole.

MYTH: Conflict between groups can be eliminated.

REALITY: Employees will always make negative comparisons between their own and other groups. The roots of potential conflict are thus always present.

MYTH: Gossip is something that just women and malcontents indulge in.

REALITY: Humans are born to gossip, men as much as women. Gossip is an essential feature of all human communities and cannot be eliminated from business.

MYTH: We are moving into an age where virtual organizations and boundaryless corporations will be the norm.

REALITY: We will develop new ways of working together, but people will always need contexts in which they can work and interact together face-to-face. Many traditional forms of organization will persist for this reason.

MYTH: We are entering the information age, in which we will work, interact, and play through virtual media.

REALITY: Although we have wonderful new possibilities for the enhancement of our experiences, we will continue to want to make and use things, interact face-to-face, and congregate with crowds in common spaces.

THE NATURE OF HUMAN NATURE

THROUGHOUT THIS book I shall be unfolding the new view of human nature that EP is giving—an integrated vision of our unreconstructed hunter-gatherer minds. It has these elements:

Motives, Thoughts, and Feelings

The world of our ancestors was full of both danger and opportunity. Life was fragile and generally quite short. We needed to navigate with skill, confidence, control, and daring. We needed to spot quickly opportunities for even slight advantage. With emotions as our radar we navigate in our new times using the old familiar ways. Now we often find our sloppy, inventive, feeling-full human impulses getting in the way of the controlled rationality needed in a complex operational environment.

Social Orientation

The most important feature of the human environment is other humans. Much of our mental hardware—plus the software programming of cultural upbringing—is devoted to making sense of one another's actions and controlling the impression we make on each other. We are designed to read minds, to interpret motives and anticipate actions. This requires empathy. This has its dangers. We can be misled. We are also adept at deceiving ourselves as a way to help us maintain convincing performances. Our goal is to build coalitions, avoid or punish enemies, and get ahead. Much of what happens in business—how we assess people, make decisions, communicate, play politics—follows instincts that work best in small communities. These instincts can backfire when we have to deal continually with strangers, as we do in the modern business world.

Cultural Instincts

Left to our own devices, we do things that try to re-create the communities of our ancestral environment: networks of no more than 150 people; hierarchical relationships; divided responsibilities; differentiated gender relations; communal arenas for interaction, display, ritual, and tournaments. We engineer opportunities for artistic expression. We tell stories and construct narratives. We look for and find leaders. These are cultural impulses. Modern business often tries to suppress or ignore them. This is the problem with much bureaucracy: too little celebration of community. The most loved businesses and business leaders play on people's responsiveness to human theater. They also provide their people with the resources and trust required to organize themselves creatively and effectively.

MANAGEMENT AS NATURE INTENDED: THE SEMLER CASE

THE BEST businesses and leaders have discovered the art of natural management, often in the midst of the most unnatural circumstances—such as under the heavy constraints of bureaucratic systems or mass production manufacturing. Let me introduce a principle that will recur in this book: what I shall call the Tolstoy principle after the opening sentences of the great Russian author's *Anna Karenina:* "All happy families are alike, but each unhappy family is unhappy after its own fashion."

So it is in business. There are a million and one ways to run a company badly, but a common set of principles underpins excellence. This amounts to a human vision of management that honors the essence of human nature. Let us look at a single case: that of Ricardo Semler, the Brazilian entrepreneur.

Several factors combined to make him ready for the experiment he was about to embark upon. First, he found himself thrust into a leadership position at a young age. This meant he escaped the trap of conventional management brainwashing, the long climb up the corporate ladder during which managers learn how to submit to norms of business rationality. While his dad was running the company, Ricardo was honing his guitar-playing skills in a rock band. Second, as the sole heir to the family business he was handed the freedom to experiment at will. Third, he was blessed with a personality and abilities ideally suited to his revolutionary destiny. His curiosity, openness to change, interpersonal gifts, and persistence combined to enable him to transform the family firm—a manufacturer of pumps and appliances—from a highly conventional business to one of the most radical and successful organizations of its time and type.

In his autobiography he describes this transformation from the time his father passed him the baton of leadership in the early 1980s. The contrast between father and son could not have been greater. Semler senior was a conservative, disciplined man who built a sound business on principles of conformity and control. From an early age Ricardo showed a different character—a creative entrepreneur who even while he was still in school, devised moneymaking and attention-getting schemes. Such early signs are very commonly visible among successful entrepreneurs. The only model of organizing that Ricardo knew well from the inside and felt comfortable with was the rock bands he loved and emulated in his youth.

He was handed the leadership at a time when the company was visibly in need of renewal. It had increasingly outdated technology, an inflexibly high cost base, antiquated management systems, and it faced new competitive pressures. Traditional organizational forms and practices were not protecting this firm in its changing environment. With a nothing-ventured-nothing-gained attitude, Ricardo took a deep breath and immediately embarked upon dramatic changes—firing many of the top managers, appointing unconventional lieutenants, and shifting the company's strategy. During this initial period of reform, the twenty-five-year-old Ricardo encountered many obstacles and drove himself to the brink of nervous breakdown through stress. But he persisted, experimenting continually with ideas from management theory.

Although his changes paid off in business returns, he wanted more. The situation didn't feel right. He was feeling increasingly frustrated that he hadn't shaken people out of the sense of malaise he sensed continued to permeate the firm. The ideas and answers he'd extracted from management texts weren't doing what Semler wanted—not far enough and not fast enough. A thought lodged in his consciousness: "What if I could strip away all the artificial nonsense, all the managerial mumbo jumbo? What if we could run the business in a simpler way, a more natural way? A natural business, that's what I wanted."

Semler admits he had no plan. He just made it up as he went along, making the changes that felt right: removing time clocks, abolishing office dress codes, opening up the office layout, consulting factory workers about the color of their overalls, removing car parking privileges. This was just the start. He went on reforming until he had achieved one of the most radical and successful turnarounds in business self-management and teamwork ever undertaken. This included:

◇ Transfer of operational decisions and target setting from managers to workers

◇ Staff collectively determining their own salaries and bonuses

◇ Completely open access to company information

◇ Informality in all communications and decision-making

◇ Job security for those over fifty

◇ Women's development programs designed and run by the firm's women

◇ Breaking up larger units into small ones

On this last point Ricardo notes: "People will perform at their potential only when they know almost everyone around them, which is generally when there are no more than 150 people."

Ricardo Semler was no student of evolutionary psychology, but he might have well have been, for an identical conclusion is suggested by the discipline. By doing what came naturally he intuitively replicated aspects of the environment in which humans evolved, and in return he reaped the reward of a highly effective, profitable business and a fulfilled, motivated workforce. His experimentation ensured not just the survival of the business but a high level of business performance.

Various features that we shall examine in this book made this work. They included a human-centered leadership built around a conception of the firm as a community; a flexible hierarchy; group self-determination in decisions; the acknowledgment of the special interests of women at work; employment protection within high-trust contractual arrangements; and clan-sized business units.

One can look around and find businesses of every kind—steel rolling mills to software designers, investment bankers to direct sales cosmetics firms—that exhibit common themes of harmony and excellence. These are firms in which qualities such as trust, support, respect, and choice are felt and celebrated. Many are small firms with a family ethos, though some large firms have managed to keep the same qualities alive in pockets of their decentralized management structures.

But firms that try to swim with the tide of human nature often find themselves battered on the rocks of economic necessity. Firms fear that they will create a competitive disadvantage for themselves if they do not follow the current business norms of uniformity, economies of scale, and rigorously controlled operations. It takes an effort to be natural in the modern age, plus the kind of courage shown by Ricardo Semler in seeking alternatives. Until now, there was little intellectual or scientific support for the attempt. This is what the new vision of evolutionary psychology has to offer.

◆ 2 ◆

Managing Against the Grain of Human Nature: The Seven Deadly Syndromes

EVOLUTIONARY PSYCHOLOGY (EP) can help us understand what goes wrong in organizational life and what we can do about it. So to start our exploration let's dwell a little on the dark side of business life. EP should be able to shed light on the things that most commonly afflict working people. Here's my list of the ills that I hear people in organizations most often complaining about:

◇ Suppressed emotion and stress
◇ Disempowerment
◇ Low-trust politics
◇ Discrimination
◇ Ineffective teams
◇ Bad decisions
◇ Management by fear

Every reader will have his or her own catalog of bad memories and experiences, but I would bet that most will fall in one or another of these categories. Some organizations exhibit more than

39

one of these seven syndromes. A friend of mine who works in a public service organization read this list and remarked ruefully, "We've got all of those at my place!"

Before we look at each in more detail, here's an EP view of the history that led us to so many maladjusted ways of working.

A SHORT HISTORY OF WORK

THERE HAVE been four ages of organizing and managing.

The First Age.   Our longest age has been the hundreds of millennia we spent as seminomadic hunter-gatherers, the period that shaped our genetic identity. Life may have been short and perilous for our kind, but it was probably also quite easy in other ways. As the smartest creatures on the planet, we developed a lifestyle involving modest amounts of work and a lot of hanging out. A little hunting and gathering on lush, well-stocked savanna in times of plenty would go a long way.

The Second Age.   Ten thousand years ago, humankind entered the era of the agrarian settlement. In such settlements, life was still clan based and communitarian. Work and home life were interwoven, but people were now never far from the oppression of kingship, servitude, and warfare. Slave states, military dictatorships, and warrior clans were the predominant political structures. Work under the whip could be terribly hard for those in slavery, though the average peasant under the more moderate bondage of serfdom would have had an easier time. Life in the Second Age was not comfortable—people often lived at bare subsistence levels—but the intensity of labor increased at certain points in the cycle of the seasons, so there was a lot of fairly idle waiting time in between.

The Third Age.   Innovations in agriculture, production technology, and transportation began rapidly to transform the world of work. Farming and manufacture became more specialized. Colonialism and mercantilism increased the mobility of people, materials, and ideas. The military and the monasteries pioneered modern systems of organizing and managing. Work and home became separate in a new age of wage labor. A new kind of economic servitude for the masses required people with jobs to work harder and longer than any previous generation. As the distance increased from our ancestral origins, so grew our psychological and social disorders—crime, social conflict, mental illness, deviance, and alienation.

The Fourth Age.   We are now entering a new age, based on a World Wide Web of knowledge and information. This can liberate us in some areas, allowing us to throw off the tyranny of the machine and find new ways of working. This is happening in pockets, but the realization of the dream is patchy, even in Silicon Valley. There is a dark side to this age, with new kinds of oppression through information and surveillance. Moreover, despite all the hype, the world of networked organizations and virtual management is distant from the mass of the world's workers and is likely to remain so for a long time.

In fact, at this present moment all four ages coexist on our planet. Remnants of the First Age hunter-gatherer peoples can still be found in its remotest pockets. Some poorer nations remain in the tribal and feudal Second Age. Elsewhere knowledge-based employment has grown enormously, but the bulk of business remains solidly in the Third Age. The control centers—the administration of modern businesses—may be increasingly wired Fourth Age fashion, but beneath this head the body is typically bound in supply chains and production systems that rely on traditional systems and methods. Commentators have prematurely pronounced the death of jobs and conventional organizations.

But change is coming to many aspects of business life, and we must be its agents, not its victims, as we have been for much of the Second and Third Ages. We are in position to take greater control of our destiny through self-knowledge than we have in the past.

In the First Age our way of life might have been hard, but it was self-regulating—a natural bridge between our genes and our environment. The Second Age saw runaway disparities in power. The few were able to enslave the many, but even in this age people retained roots in stable communal structures. The Third Age has seen a massive erosion of these structures. Removed from large parts of the developed world have been the extended family, communal ownership and organization, local self-government. As we enter the Fourth Age, many of us live in unprecedented affluence, able to avail ourselves of a staggering array of consumables and entertainments. Accompanying this affluence is a rising tide of crime, drug abuse, violence, pornography, and mental illness. Psychological fulfillment from religious faiths has become elusive. Many people are victims of twenty-first-century ennui, the bored emptiness of the perpetual consumer.

This is the price of the widening misfit between what we are and what we have created. The Seven Deadly Syndromes are symptoms of this misfit.

SYNDROME 1. SUPPRESSED EMOTION AND STRESS: THE TYRANNY OF RATIONALITY IN THE HIGH-PRESSURE ORGANIZATION

Dave is a member of a telesales team for a replacement window company. The business is highly competitive, and his pay is 50 percent commission on the sales he makes. He has a young family, and frequently he works late to make his quota. The local economy is experiencing a downturn, and demand is declining. He often has to endure abuse from people he has cold-called on the telephone. His calls, like those of the other operators, are monitored by a supervisor. He has seen several colleagues let go when they failed to make target or their manner of dealing with customers was judged to be poor. Dave's wife is expecting their third child, and there are health complications. This is stretching his finances and his emotions. One day at work he has a sharp exchange with a supervisor about a call. The supervisor warns him to get it together or he'll be out of a job. Dave is losing sleep with worry. He doesn't want to burden his wife with his problems. He tries to keep up a brave front.

Lots of things seem to be awry here, but this tale is unexceptional, especially in highly developed economies, where people are working under increasing pressure. Work and home are sealed off from each other. The nuclear family is isolated and unsupported. No allowance is made at work for people's life circumstances. Managers command and control through reward and punishment. Errors and failure are not tolerated. The expression of emotion is strictly taboo in the work environment.

Stress is fast becoming the number one hazard to the health of working people, with alarming increases in the incidence of every kind of neurosis and psychosis. Two factors are to blame. One is the increase in pressure upon people from all sides. The other is the withering of support from cohesive community or family. These effects come together at the leading edge of the market economy, where relentless competition drives firms hardest. They are often made worse by the companies' location in rootless, anonymous, and transient urban populations. We wonderfully adaptive humans tolerate the misfit of such "hyperdynamic" firms because if you can stand the pace the rewards they offer are significant. People within such firms—like soldiers going to battle—become fiercely avoidant of emotional expression, especially those that signal weakness. Loss aversion makes people fear emotional disturbance like a contagious disease.

Failure-phobia is nearly universal in Western cultures but is especially prevalent in corporate cultures where competitive pressure enforces "zero tolerance for errors" (a telling slogan of the quality movement).

In healthy communities, loss and failure are openly acknowledged. They allow people to share responsibility for each other's feelings. In tribal communities, rituals help frame emotions and expunge shame. Free from the emotional burden, a sinner or loser has a genuine chance to learn from his or her mistakes. Many of the environments we live in contain no slack for this kind of indulgence. Stress is especially prevalent for the staff dealing with customers in some service businesses. We would all like to be rid of the surly waiter who unsmilingly slams your food down on the table in front of you, but from the point of view of the service provider the "have a nice day" ethos has a cost. This is called emotional labor—occupations where you have to wear a mask all day long. In telesales, operators are trained "to put a smile into the voice." And airline cabin crew members must maintain unlimited sweet reasonableness in the face of encounters with some of the world's most obnoxious travelers.

Something has to give if organizational cultures exhibit a studied blindness to emotions and give zero support to staff working under conditions of high demand. The accumulated poison of suppressed negative emotion seeps through the cracks. Office staff get sick. Leaders drop dead from heart attacks while rushing from meeting to meeting. Managers fly off the handle at any provocation.

We must reject the view, held not just by working people trying to get a job done but also by some professional social scientists, that emotions are events which interrupt normal thought. Daniel Goleman has popularized the concept of emotional intelligence as a counter to the emotion-as-disturbance view. Far from being an aberration, an awareness and acceptance of emotions mark the superior leader and effective performer. People who lack this sensitivity can cope with difficulties only within a narrow range of human reactions. Neuropsychology is telling us that emotions are not just background for the voice of reason. In our neural circuitry they are primary. Every fleeting experience passes directly through neural emotional screening centers. They notice and react to anything out of the ordinary and put us on guard against danger. Our feelings set the agenda for all our so-called rationality. They switch us on to what makes us feel good and turn us away from what makes us feel bad. They tell us what to pay attention to. They filter and stimulate our thoughts. They guide our reactions to events even before we have had time to stop and think about them. Our emotions are our radar in an uncertain world.

SYNDROME 2. DISEMPOWERMENT:
QUALITIES OF LIFE IN THE BELLY OF THE MACHINE

Jo is in the claims adjustment section of a large insurance company. When she joined the firm, she used to deal with customers directly and, under supervision, reach independent decisions about claims. Recently the operation was reengineered with the installation of a new computer system. Now her job has narrowed. She has to extract data from claims forms and enter them into the system; then she has to follow what it instructs her to do about them. A separate customer service section deals with all claimant calls. She enjoys the company of her colleagues, but most of the day she sits in isolation at her workstation. Both she and her friends at work feel no affection for the company or the job. Absenteeism is high, and Jo herself often feels too low spirited to show up for work.

This is a story about bureaucracy and its disempowering effects, especially at lower levels. We have great ambivalence about bureaucracy. We use the word *bureaucrat* as a term of contempt, yet we depend on what bureaucracy delivers. It is among our greatest human inventions—a way of organizing that has accomplished vast projects, won wars, and provided many of our most needed commodities and services. Yet we also see how it can crush the human spirit and make monsters of the nicest people.

Bureaucracy coordinates people through hierarchy, division of labor, rules and procedures. When it functions well—and even when it operates badly—a bureaucracy gets through a lot of work economically, like a big machine. People are substitutable cogs in its innards. That's a large part of its attraction. In theory, by over-riding the human element you minimize operational risks. You become immune to bad leadership. Everyone is replaceable. Roles rule. Control is the byword, from top to bottom. Because there is a limit to the number of workers who can be directly controlled by one person, a small army of supervisors is needed. Another layer is needed to supervise the supervisors and so on until you reach the top level, creating the classic organizational pyra-mid. At the height of its success in the mid-1980s, IBM managed an empire of around half a million people through just such a linked hierarchy.

In bureaucracies, technical innovation is used to remove the need for skill. Mechanistic control is always favored over human fallibility. Jo's story illustrates this, and it continues to be true in every industry from car production to retail banking. The jobs that can't be mechanized are often not fit for humans but our adaptability allows us to submit to them. Karl Marx invented the term *alienation* to describe the corrosive effect of the factory sys-tem on social solidarity, though this did not prevent Communist states from enthusiastically adopting it to secure its economic benefits. Henry Ford realized the risks and sugared the pill of life on the assembly line by giving workers above-average pay and fringe benefits.

Alienation means four things, all of which run against the grain of human nature. One, you have no power in the social group. Two, you are isolated from other people. Three, the tasks you are asked to do seem meaningless. And four, you have no sense of being part of a community. No wonder unions became powerful in traditional bureaucracies—they offered a refuge from alienation. Workers could recapture some personal influence, enjoy associating with others, engage in meaningful actions, and regain a sense of community—as well as the hope of a greater share of the rewards.

Quite apart from the economic benefits, our instincts make us embrace bureaucracy for another reason: love of status. We create elaborate hierarchies because they are also compelling motivators. A long career ladder can keep a disempowered person striving on a steady drip of increments. The threat of loss of position adds deterrent bite to the system.

Many species have pecking orders, and in social animals jockeying for position takes a lot of time and energy. Why? Because it is pretty universal that those nearest the top get the best mating opportunities, the most resources, and the best lifestyle. The drive for status is evolution's way of ensuring that the best genes, as well as the boost of resource-rich parents, get passed on to offspring, giving them the best material start in life. Conversely, in all social species it is bad news to be at or near the bottom of the heap. Life expectancy is shortest. Stress is greatest. Risks are highest.

Humans are in exactly the same territory. Life for people at the bottom is completely different from life for those at the top. With high status come material rewards, influence, and well-being. At the bottom are meaningless work, meager rewards, and no power. Worse, at lower levels you get sick more, you have a more impoverished social life, and you die earlier. Controlled medical surveys in one huge investigation of civil servants found the bloodstreams of low-status staff awash with chemicals associated with high stress. This was due not to poor diet or any other objective factor but is a pure status effect. Controlled studies of captive monkey colonies have found exactly the same pattern:

concentrations of the same chemicals flowing through the veins of low-status animals.

SYNDROME 3. LOW-TRUST POLITICS: CONTRACT VIOLATION IN THE POLITICAL ORGANIZATION

Andrea graduated from college and joined a large oil company as a trainee in the shipping department. She did well during her first years, successfully managing a number of small projects. Her boss gave her favorable ratings and hinted during one of her appraisals that she could find herself on a fast-track development scheme before long. Shortly afterward, her boss was transferred without warning to another divisional office. Her new boss has also given her positive ratings but has made no promises about future prospects. Andrea learns that two of her peers, who happen to be male and joined the firm at the same time that she did, are to be promoted out into other divisions. Andrea is piqued. She knows they have achieved far less than she has in their time with the company. She thinks they have benefited from working closely with the new boss on his pet project. When she expresses her dismay to a colleague in the same section, who has been with the company a little longer than she has and has also been passed over, he laughs cynically: "You should learn the rules around here. If you want to get on, you have to spend more time cultivating relationships. You'll never get yourself promoted unless you can get yourself invited to Brad's [the new boss's] Sunday brunches."

If you escape the cryogenic chambers of bureaucracy—getting frozen in a regimented hierarchy—you might find yourself having to survive the heat in Andrea's world: the political organization. Actually, many people in bureaucracies complain about the politics involved, and with some justice, for the human instinct for politics—maneuvering to advance one's interests—is irrepressible. Because in a bureaucracy not everything can be nailed down by rules and standard operating procedures, and people furiously play politics in what little area there is left to do so: fighting madly for the biggest office, trying desperately to get noticed, playacting at leadership by fiddling with administrative trivia.

In full-blown political organizations, such as many commercial corporations, the competitive impulse has been allowed, even encouraged, to slip the leash altogether and everyone is playing games for real. The last two decades have seen a steady shift from the bureaucratic toward the political model as companies have sought greater flexibility, especially in human resources practices. Inflexible career and reward systems have been replaced by merit-based performance management and self-managed career development. Under conditions of low job security this is a double whammy. It warns that you will be vulnerable to the subjective judgments of those more powerful than you but that no one but you will take responsibility for what happens to you. This might be welcome under conditions of security and high trust but in a low-trust environment you will be fighting for your life.

Three key features of the human design create this problem. One is loss aversion, the fear of losing social position. Second is sensitivity to contract violations. Third is our desire to manage social impressions. The last two features stem from our origins as clan-dwelling animals. We need to monitor and control our standing in the community. This is essential for survival, since in the community of the clan it mattered a great deal who could be counted as trustworthy. Business leaders all too often underestimate their employees' sensitivity to implicit contracts, how easy it is to break such contracts, and how lasting the damage is if people think they have been broken. The climate of an entire firm changes dramatically when people are unexpectedly laid off, even when the layoffs directly affect only a tiny proportion of employees. Trust, like a tree, takes a long time to grow but can be cut down fast. In business, people are often unhappy about failures in perceived distributive justice—unequal shares of rewards and opportunities—but they get much more upset about breakdowns in procedural justice, the thought that *the way* decisions are made is fundamentally unfair. In the political organization a lack of either kind of justice bedevils promotion decisions, budget allocations, pay and fringe benefits, and cost reduction practices.

The political organization also calls upon our natural skill in impression management, the art of presenting ourselves in a favorable light. We are equally sensitive to the self-presentational skills of others. In the world before work organization, social cohesion was a primary goal, and the skills for maintaining it were highly valued as vital to the survival of the individual and the welfare of the group. In modern business, these gifts are both a help and a hindrance. They are the stock in trade of every salesperson and negotiator doing deals, but they get in the way when you try to make impartial objective judgments about people. Appraisal systems are designed to restrain our hardwired tendency to rush to judgment about individuals. It is said that an untrained interviewer forms a lasting impression of a candidate in the first few minutes of the interview. It requires training and structured methods to make interviews scientifically reliable. But even then, interviews remain highly vulnerable to the chemistry of instinctive affinity. It's fairly easy to render objective judgment about someone's technical skills but once we slip into the world of intangibles—such abstractions as leadership qualities and interpersonal skills—objectivity becomes well nigh impossible.

Some organizations are almost entirely political. Everyone is intent on playing the game of how to get ahead by being plugged into the right network, being liked by the right people, and projecting the right image. You will find this happening most where the outputs of the organization are intangible and hard to measure—for example, in education, some areas of health care, and many voluntary or not-for-profit organizations. Sometimes the politics are highly male: games of competition and dominance.

Sometimes the politics are female: games of acceptance and moral superiority.

Political organizations can work well when the politics lie on the surface where everyone can see them, and where the values they represent are shared and important to the group—that is to say, something akin to the true community spirit of the clan. More often, though, in contemporary society we lack the vital preconditions—trust and a sense of shared fate. Most political organizations are full of uncertainty and mistrust.

SYNDROME 4. DISCRIMINATION: TRIBALISM AND WARFARE IN THE SEGMENTED ORGANIZATION

Rosita is a medical administrator. Born of Puerto Rican immigrants in a poor area of New York, she married a sales executive. She worked for a few years in a city hospital but when her husband was transferred to the Midwest, she got a job in a private medical center there. The center is a large facility providing hospital treatment and a range of community health functions for a largely middle- to upper-class population. After a few weeks on the job, she comes home from work in a state of high emotion. She explodes to her husband: "I'm fed up with this place. Everybody is trying to protect their little empires. The nurses keep information from the doctors so they can work without interference. The doctors don't cooperate with each other but try to hand off to each other the cases they don't want. The technicians work as they please and despise everybody else. They all regard the administration as their enemies and try to avoid doing any paperwork that is requested. I am fed up with seeing patients suffer because these people won't work together. To cap it all off, I overheard someone referring to me as 'that wetback in administration.' It doesn't help not being local. Everyone's so parochial and small-minded."

Rosita has encountered the ugly side of tribalism. Curiously, it crops up most often where people are rooted in traditional communities. Only in the anonymous individualism of a cosmopolitan city are its effects muted, and even here within the ranks of large and small corporations you can see its effects. It is astonishingly easy to induce tribal feelings. Imagine this:

You are a schoolteacher. One day you walk into your classroom and, completely at random, you assign half your pupils to the blue group and half to the red group. Now you get them to do some activity in their separate groups, after which you ask each group how they reckon they are doing compared with the other group. In the absence of any objective evidence, each group will claim its superiority over the other. Ask group members if they have any opinions about the individuals in the other group. Each group will be quite willing to make disparaging remarks about what sort of people are in the other group, even if their best friends are split between the groups because of your arbitrary allocation. Now get the two groups to engage in some mildly competitive game and watch what happens. Positive attitudes toward one's own group and negative attitudes toward the other will start to escalate. Hostility spills out beyond the classroom. People develop new friendships that do not to cross the group boundary. We enter the world of feuding gangs, of *Romeo and Juliet's* Montagues and Capulets. You decide to end the experiment and bring them back together in class, but you see the reds and the blues continue to segregate themselves. It takes several days of activities to integrate across the boundary you have created for things to get back to normal.

Now imagine that you didn't group your students randomly but used some criterion such as what things they are good at or which side of town they come from. Suppose it was some visible attribute like whether they were thinner or fatter than average or had dark or light hair color. When you question your two groups about each other, don't be surprised if some come up with prejudicial statements that prove to be quite popular with their pals—

for example, a blondie says, "I think people with dark hair are weaker than people with light hair."

Let's leave the schoolroom and enter the multidivisional international corporation: Procter & Gamble, the Sony Corporation, or General Motors. Here you will find groupings of all kinds—by hierarchical level, by job function, by location even within buildings. In some areas, you will find groupings by sex: concentrations of women in human resources or accounting, perhaps. In other areas you will find that racial or national groups separated out—immigrant labor in manual operations, for example. Other groups can be distinguished by their educational background—for example, the techies in R&D or the liberal arts types in marketing. In many international companies you will also find elite groups—fast-tracking recent college recruits being groomed for distinction as a separate cadre of "global managers." They are given a succession of international assignments to polish their manners and their résumés—the modern business equivalent of the Victorian Grand Tour. They never stay anywhere long enough to make a serious mistake or to put down roots. They are "endowed" with a network. These groomed stars are often resented by others in the firm, who rightly see this as a caste system, sustained by self-fulfilling investments of management attention.

One way or another, the scope for tribalism in large businesses is very great indeed.

In 1997, three British executives won a suit for unfair dismissal by a U.K.-based Japanese financial information house. The executives said they were treated as second-class citizens, as secondary and expendable. Others reported that *gaijin* (foreigners) faced aggression and exclusion. They were also paid lower on a two-tier salary structure and were severely restricted in their decision-making powers. Secretaries were described as part of a secret police force; in the words of a Japanese manager, "They are listening posts—because they listen to what you say and report back to us."

This kind of tribalism has its roots in three hardwired predispositions. The first is a powerful group orientation. We are programmed to identify with the groups we are members of. Our sense of identity is tied to the need to belong, a key factor sustaining family and clan life. Second, we are hardwired to think in categorical terms, to love lists and think in terms of taxonomic groupings. This is a product of the systems that helped us to navigate around a world where some plants are inedible, some animals are dangerous, and some people are untrustworthy. Third is our interest in status. These three tendencies create the basis of social discrimination, the tendency to perceive differences between groups and to make in-group, out-group discriminations. Humans routinely take pride in their own group and devalue the members of other groups. It helps us feel good about ourselves and our social position to think well of the groups we belong to. This has various consequences: suspicion toward strangers, resentment of newcomers, dismissive attitudes toward other groups. The need to belong to the right group is heightened where there are powerful pressures to achieve, as in the Japanese firm. Belonging to the right group has always been important where there is competition for scarce resources.

A fourth factor in our makeup gives focus to our discriminations. This is the fact that much of our brain is taken up by our prodigiously powerful visual system. Our very language reflects this. It is full of vision words (do you *see* the point I am making?). The consequence is that the more visible the characteristic, the more it is likely to be a hook to hang discrimination on. Skin color is highly visible and consequently is a basis for social categorization in just about every culture.

Personnel selection systems unconsciously play to tribalistic impulses. The exhaustive interviews, tests, and other devices in recruitment are not pure science. They contain a strong element of cultural chest beating, sending the message: "Our selection is tougher than our competitors' therefore we are a better organization to belong to." This does not mean that we should abandon elaborate recruitment methods, only that we should regard with healthy skepticism their claims about what they do and separate in our minds their operational from their cultural functions.

SYNDROME 5: INEFFECTIVE TEAMS:
WEAKNESS AND FAILURE IN GROUP WORKING

Arthur is a recently retired army general. He has been asked to serve as a nonexecutive director on the board of a company following its initial public offering. The head of this highly successful family engineering business is a man who served under Arthur some years previously. Arthur accepts the job but after a few meetings apologizes to his friend that he will have to withdraw from the board for personal reasons. The real reason is that after a lifetime in the military, where meetings were short and to the point, Arthur feels intensely frustrated and impotent with his role on the board. Its fourteen members spend meetings poring over trivia and arguing inconclusively about broad strategic questions. Any important decisions have already been made by the firm's executives, and the board simply rubber-stamps them. No one on the board seems concerned about this. The company is turning in healthy profits, and Arthur gets the feeling that the main reason others attend is to network with each other. The conversations members have before and after meetings are more animated than what takes place during the board's formal deliberations. Arthur's attempts to get the others to focus on actions are listened to politely but ignored. He feels they have developed a comfortable routine and he is an unwanted and interfering outsider.

Arthur has stumbled into one of the most puzzling contradictions of business life—why we, as creatures designed to cooperate and work in groups, so often do it badly in organizations? Fortunately, there are lots of counterexamples of inspired teamworking in all walks of life, including the military. Great groups also can be found in theater, music, and the arts; in science and technology; in sports and many areas of business. It is the excellence of many project teams that has been responsible for some of our greatest inventions.

These examples show that groups work best when they can organize themselves around things that really matter to them and their wider community. The hunting party, the craft-working group, and the performing arts troupe are examples from the life of clan-dwelling peoples. We are equipped with skills appropriate to just this kind of cooperation: empathy, speech, the desire to share and trade contributions. These have enabled our species to survive. If you ask people what is characteristic of the best groups they have ever been members of, the same traits keep being mentioned: small, informal, egalitarian, inspired by shared goals, the knowledge that what they do is recognized and valued. They report that the group does its work in a spirit of energized sharing, with much constructive self-examination. Individual members seem to know exactly what their distinctive contribution is, but at the same time they have a feeling of "losing" their egos in the group. The group takes on a life of its own.

Groups fail in organizations mainly because they are *in* organizations. They reflect the surrounding cultures—rigid, formal, and highly political in many cases. Arthur's board exhibits some of the classic problems:

It is too large. Fourteen is unwieldy for genuine teamwork. We feel at home and function best in family-sized groups of five to nine members.

It is also too formal and bureaucratic. The rules by which it is run exist for statutory rather than functional reasons.

Its goals are unclear and trivial. Real decisions are made elsewhere (by the company executives).

Groups have to focus around identifiable objectives that matter to their members. Arthur's board is not so exceptional. Many company boards are not involved in real decision making and exist only in order to satisfy the requirements of stockholder governance. At every level in many companies, you can find so-called teams that are unnecessary: committees and working parties set up for political or faddish reasons. To be a member of one of these is demoralizing when you realize that no one outside the group really cares about what you are trying to do or is interested in what you produce. Another problem typical of big companies is people acting as representatives of their local tribe—marketing, finance, production, etc.—rather than immersing themselves wholeheartedly in the team and its mission.

We need more good teams in business, but people are reluctant to join them or set them up because they have already suffered from too much exposure to useless committees. What we need are teams that pursue clear goals on behalf of the firm as a community, transcending functional divisions and hierarchical levels. This is what happens naturally in small firms. Large firms need to try to capture the spirit of small companies by building a team culture at all levels. Sadly, many of the ways large corporations organize themselves create conditions that make this almost impossible to achieve.

SYNDROME 6. BAD DECISIONS:
HARDWIRED FOR CROOKED THINKING

Arlene, a buyer for one retail clothing firm, has been recently recruited by an executive search (headhunter) agency to work for a large competing company. Shortly after joining she is appalled to find out what bad shape this company is in. Accurate performance figures have largely been concealed from the outside world, but Arlene suspects this cannot last for long. The problem is weak leadership at the top, leading to department heads competing with one another for resources. As a result of these turf wars company executives have made a series of poor decisions. The head of buying switched to a cheaper supplier that subsequently went bankrupt, forcing the company to find new supply sources at far higher than the original cost. The production department bought new cutting equipment from a foreign supplier, but the new equipment keeps breaking down because those who operate it are inadequately prepared and trained. The marketing department embarked on an ill-fated advertising campaign to try to boost sales in existing clothing lines while competitors were developing new products. Decisions in every area seem ad hoc and unregulated.

Arlene is the victim of two phenomena we can call *migrants' optimism* and *residents' myopia*. These are human biases that help us adapt to change and find stability in changing environments. Migrants' optimism—the tendency to think that the grass will be greener at one's next stopping place—is confidence building. It is a piece of self-deception that helps insulate mobile people whose lives consist of movement from known to unknown environments in order to survive. People who move from job to job are typically vulnerable to the illusion that the organization they are joining will be free of all the human foibles and inefficiencies that characterized their last stopping place. Young people taking their first full-time job after college are especially prone to migrants' optimism. They do not even have the benefit of experience to counter the illusion, and they are flush with youthful ambition.

Residents' myopia is the tendency of the long-term denizens of a subculture to become blind to the inefficiencies, irritants, and foibles of their local world. This too is adaptive, a force for stability in a community. It makes people more empathetic and forgiving of one another and also of themselves.

This has been a source of hair-tearing frustration to many a new manager who enters a firm to be shocked at what the inmates take for granted. One reason companies use consultants is to get the benefits of newcomers' insight without having to hire them permanently. Migrants between cultures notice things that locals don't see, an obvious benefit of using expatriate managers. But it has a sting in the tail when the expatriates come home after their tour of duty. They are typical victims of migrants' optimism, remembering only the good things about home, so when they return they become irritated and depressed, not just by all the annoyances they had forgotten about but also by minor screwups: irritations that never bothered them before but now seem bad to them.

How can such smart creatures be so stupid when it comes to making decisions? Why do we seem to get so tied up with petty matters, make ill-considered bets on chancy outcomes, and let ourselves get carried away by momentary impulses? The business literature is littered with major and minor screwups: ill-prepared product launches, inept advertising strategies, poor selections, engineering system failures, and major accidents.

Although we are undoubtedly the smartest creatures yet to appear on this planet, we have severe limitations when it comes to rational thinking. There is no reason to clutter up our brains with computational capability when we have more pressing matters to attend to, such as getting mates, finding food, avoiding danger, and striving to enhance our standing.

Many minor and major disasters come from the sheer complexity of the environments we are designing and our inability to keep pace with them. Evolution has a bias toward economy and simplification. The giddy acceleration of technical innovation through information technology, biochemistry, materials science, and many other disciplines increasingly outruns our limited capacities. Since the invention of working machines we have been aware of the problem and try to compensate by designing human-proof operational and control systems.

Many human errors come from a more purposeful kind of incompetence.

We have inbuilt biases to what we notice and remember, being drawn to whatever is shocking, familiar, and emotional.

Our reasoning is faulty: we need paper and pencil to figure out elementary logic problems.

We calculate probabilities badly. Evolution did not give us calculating machines for brains. It has been much more useful to be quick witted and intuitive, tuned in to people and events rather than numbers and facts.

Allied to these biases are the confidence and optimism biases that help us, in the words of Shakespeare, "summon up the blood" for the difficult, dangerous, and unknown. They also have a social effect. Others around us sense and share the feeling. The confidence of the leader infects the followers, a major element in what people call charisma.

Finally, what Arlene also discovered is the political processes surrounding bad decisions and accidents and the powerful motives to conceal, protect reputations, and distort with hindsight. Loss aversion plus our communal instincts incite everyday conspiracies—all the more so if under the arc-lights of public scrutiny.

Many companies proclaim the value of being learning organizations. In many firms there is a gap between rhetoric and reality.

SYNDROME 7. MANAGEMENT BY FEAR: PATHOLOGIES OF PUNISHMENT AND PARANOIA

Randy has been a junior manager in the human resources department of a large bank. Feeling frustrated at the lack of opportunities for advancement, he takes a job as the HR manager in a medium-sized firm that produces office equipment. The post has been newly created. During his initial interview he is impressed by the forthright and forceful style of his new boss, the firm's founder and owner. But soon after he joins the company he discovers that everyone is terrified of this man, a bully prone to violent mood swings. One day the boss is all beaming affability; the next he seems angry at the world. Frequently he can be heard bawling someone out in his office. People tiptoe around him with trepidation. Randy tries to get him to talk about succession planning for key management jobs. The boss dismisses the idea. "Just bring me the list every year," he says. "I'll tell you who's to be promoted. You stick to payroll and training." At year-end the boss hands out bonuses to his favorites. Randy notices that people of real quality tend not to stay long but leave for other companies. Managers promoted to senior positions typically share the boss's authoritarian attitudes and style but lack his charisma. Everybody else is stuck, unpromotable and unemployable anywhere else—they've been there too long. They just keep their heads down and try to stay out of trouble. After twelve months Randy quits to find another job elsewhere.

Randy learned his lesson: take a bit of time to find out about a company's culture before you join. Two firms in the same business can exhibit very different characters. People who stumble myopically into a firm without checking it out don't stick around too long if they find the culture not to their liking. In other words, there is a matching process between the personalities of the people who join a company and its culture. Those who are not free to quit have to develop attitudes and values that fit or suffer in silence.

Firms develop their cultures over time. In some firms the culture is strongly influenced by a production technology that forces people to work to a set method. An assembly line or a call center compels people to interact in a particular way. In firms it is "the people who make the place," and the people's motives and skills differ according to the nature of the business. This is why the climate is so different in, say, advertising, banking, and steel production.

The less bureaucratic a business, the more its leadership shapes the culture. This is what makes two outwardly similar firms quite different in their internal ambience. Many small to medium-sized enterprises are governed by old-fashioned authoritarianism—rule by fear. Like many other mammals, we are highly sensitive and responsive to dominance. But we do not have a single model of leadership because of the varied tasks and environments the human clan has to cope with. Sometimes human dominance resembles the model of chest-beating alpha males who bully their way to the top. At other times it is similar to a wolf pack, where the acknowledged leader is the animal best able to harmonize relations within the pack by neutralizing aggressive individuals and resolving conflicts.

The world of business seems to contain many more chest beaters than conflict resolvers. Competition, ever present in hierarchies, results in tough fighters who succeed by defining the world as threatening. If they can get the same perspective into the heads of their followers it will shore up their position. This is often not hard to do. It is easy to play on people's natural loss aversion. Once this happens, paranoid people with an aggressive style become favored at lower levels, and a culture of fear spreads throughout the firm. This may prove successful in the short term but ultimately contains the seeds of its own destruction. Despotic businesses are unstable, often rocked by crises, and unable to hold on to key people.

I remember seeing this in a textile factory run by a powerful self-made man. Everyone was afraid to bring him their problems. They had seen too many bearers of bad news decapitated. Instead they tried to paper over problems rather than admit what might be perceived as failure. Making a mistake often meant public humiliation. At one board meeting I watched this boss openly ridicule one of his key executives over some minor omission. What was most disturbing about this was that no one, apart from me, seemed at all embarrassed. In this firm, the episode was just business as usual. Some months later, in a state of near collapse, the firm was taken over.

Violence is close to the surface in many firms, not just because of the culture but due to the random uncontrolled impulses of individuals, mainly men, who have a powerful yet thwarted desire for dominance. Bullying and harassment are frequently reported in the media and are routinely the subject of legal actions in North America and Europe. I have frequently heard male managers use the metaphors of combat: finance traders talking about killing the invisible enemy through their computer terminals and managers talking about the execution of some unfortunate colleague for failure. Men especially are susceptible to the lure of dominance; they like to feel strong, brave, and in command. Fortunately people differ in their need for these feelings, and being domineering is not the only way to exercise effective authority and satisfy one's power needs.

TOWARD SEVEN SOLUTIONS: THE EP WAY

LET US not idealize our ancestral environment or perpetuate the myth of the noble savage. Our Eden of hunting and gathering was not a mixture of safari-based bonding for the boys, homely craftwork and child care for the girls, and sweet harmony for all around the campfire. Human society has always involved the Seven Deadly Syndromes: emotional distress, low-status blues, cheating, intergroup warfare, intragroup failure, bad decisions, and despotism. However, in hunter-gatherer societies these ills are regulated by the community and counterbalanced by the strong forces binding people together in kinship and cooperation.

In contemporary organizations we need to rebalance these elements. It is not hard to identify the conditions under which the syndromes flourish: rigid hierarchy, narrow membership profile, inflexible technology, extreme work intensity, external threat, high membership turnover, segregation of the sexes, and scarce resources.

Take the phenomenon of overwork: people working longer and longer hours, allowing work to eat into their leisure time, attending evening meetings, and taking piles of work home. This is becoming a disease of epidemic proportions in many businesses. And it is unnecessary—we could work smarter instead of longer. By dint of some intelligent time management, most people could do their jobs in fewer and more reasonable hours. So why doesn't this happen? It is caused by a mix of the elements we have considered:

◇ High volume of work plus inhibited expression of discontent
◇ People defending themselves against the threat of job loss in an environment of fear
◇ People working ridiculously long hours to look good in highly competitive climates
◇ Machismo bosses setting tough expectations to assert their dominance

So what might the evolutionary perspective suggest we should try to do to avoid the worst effects of the Seven Deadly Syndromes?

1. *Honor emotion.* Let's accept the importance and unavoidability of emotion at work, raise awareness of people's emotional experience, and develop ways of helping people work through rather than against their feelings. How can we do this? It is a question of company culture. People from the top down must be allowed to be open about their feelings, and managers who display emotional intelligence must be promoted.

2. *Flex the hierarchy.* You can't eliminate hierarchy, but you can make it less rigid by multitasking. There will always be people in the lowest positions. Let's create healthier organizational cultures by protecting and supporting these people and by making them feel more included in the business community. Inclusive management practices are important

here—for example, project teams in which lower-ranking members have as much say as more senior people.

3. *Keep politics on the surface.* Ensure that justice doesn't get corrupted by politics. All businesses are political, but let's keep the politics in the open where we can see and manage them. This means frankly discussing how people's interests differ and not trying to smooth out diversity. Openness is needed about how people can network to find new career tracks. Management has to be more up front about the real nature of the psychological contract—that is to say, what people can really expect of the company and what it can legitimately ask of them.

4. *Keep organizations fluid.* Companies should engineer job moves not just for key people but for other staff, to take them across the Chinese walls dividing a business—such as those that often exist between sales and marketing or production and design. New liaison roles and projects can be developed around the key objectives of the business. Groups whose interests diverge can be creatively brought together to explore the goals that might unite them.

5. *Make groups organically diverse.* Organizational units need to be kept small. We need to maintain diversity of group membership and ensure that groups function informally, flexibly, and nonhierarchically. What groups produce must have real influence within the wider community. Empowered and self-managed teams have a great track record in some businesses. In others they fail because top-down empowerment is halfhearted.

6. *Use systems that counter human biases.* We should develop and use decision-making aids that protect against the irrational tendencies of human thought and judgment. End users and operators should be encouraged to question taken-for-granted decision processes. Leaders can encourage people to

be open about errors and learn from them. People who ask for help should be rewarded, not punished.

7. *Devalue dominance.* Encourage exposure of the abuse of power and look for ways to keep it in check. Executives need to consider whether prospective leaders can work as team members with their staff. The temptation is to select powerful, impressive leaders. Dominance alone is a narrow, maladaptive strategy in the long run. Other qualities are needed.

None of these is easy. There are major economic and psychological barriers to achieving this wish list. Short-term gains of efficiency or personal gratification stand in the way of change that might yield longer-term benefits. Many of the barriers are structural. We have adopted ways of working that allow us to extract the most resources from our environment with the least effort. These have forced us into some dysfunctional patterns. Now we have new options resulting from innovations in technology, information, and communications. Using these while taking account of what we are learning about our evolved psychology can help us manage and organize with rather than against the grain of human nature.

• 3 •

Sex and Gender:
Old Models in New Roles

THE MALE ORGANIZATION

SEX IS the biological difference between men and women. Gender is the social difference between the sexes. Sex is encoded in chromosomes, brain structures, and bodily differences. Gender is encoded in symbols, manners, and beliefs.

Sex is what it has always been. Modern men and women differ now just as they did in our earliest days as a species. Sex and gender are intimately connected: gender roles change with the times in order to keep pace with the demands being made on men and women, but this change occurs within limits set by our biology. The same themes keep cropping up in gender roles, across historical eras and cultures. These are anchored in sex differences in human physiology and psychology.

The way we work and organize ourselves has always accommodated sex differences. As far as we can deduce from archaeological remains and anthropological investigations of surviving hunter-gatherer clans, in the long preagrarian first age of human history men and women lived and worked together in quite egalitarian and cooperative style. Yet at the same time the society would have been patriarchal: men in leading roles, sons inheriting fathers' status, with much of the daily labor divided by sex—hunting done by men, women doing most of the work closer to camp, including foraging.

In the agrarian second age, male power grew without restraint. Occasionally women ruled empires, but more often they wielded influence from behind rather than on the throne. Women were usually in positions of subordination to male power. Women's work was restricted mainly to domestic labor, work on the land, nursing, welfare, child care, and some arts and crafts. Men in the military, in monasteries, and in basic industries and trade controlled all the main areas of wealth creation, science, and artistic achievement.

The same arrangements have prevailed for much of our third age of industrial society; only now we have constructed elaborate systems to do society's work. These are predominantly male. In business today we find mainly:

◇ Men in leadership positions
◇ Elaborate hierarchies that feed male status aspirations
◇ Competitive systems enabling men to test themselves against one another
◇ Goal-focused strategies and procedures to satisfy male achievement motivation
◇ Highly specified tasks and roles that fit male single-mindedness

Although women have been increasingly admitted to this world, first to the lower strata to satisfy the booming demand for industrial production and more recently to positions of management and professional responsibility, *the world of business organizations remains male in rationale, design, and functioning.* There are predominantly female-staffed organizations in health, education, and welfare, but even these are organized and run on the male model, with men continuing to command the top slots. We shall explore why this is and whether it is an inevitable legacy of our evolution or whether with the aid of new kinds of organization we can enter a new age. Can we move beyond the traditional male organization toward a viable alternative female one? More than this, in the fourth age of postindustrial society can we create gender-neutral businesses?

The answer is two-sided. Yes, new ways of organizing can give greater scope to female preferences and will be all the more effective for doing so. But no, we cannot create gender-neutral businesses, because even when men and women want to do the same things their approach often calls upon distinctive styles and brings them different satisfactions. Traditional gender ordering persists when these biases are allied to the way we organize—the male model of control is inherent in the economic rationality that drives for profitability, cost reduction, quality control, and efficiency. However, the information age does offer us fresh possibilities for coming back closer to our evolutionary prototype of a more flexible interchange in the contributions and responsibilities of men and women. Yet the disturbing possibility remains that as long as the profit motive drives how we organize, the more ruthless push of male values will continue to predominate.

THE GENDERED WORLD

You are on a visit to the R&D facility of a large computer firm. You stop at the entrance, where two female security guards are swapping jokes with each other as they lounge against the barrier by the gatehouse. One disengages herself momentarily from the conversation, takes the cigarette out of her mouth, and points out where you should park. You find the R&D block, where a smartly dressed young man from reception greets you and walks you to the director's office. A silver-haired man in his late forties introduces himself as the director's PA and says you should go right into the meeting room. Seated around a table are the twelve computer scientists of the facility. At the head of the table is a gray-haired woman, the director. Three of the other four managers in the group, the older and more senior scientists, are women. The other seven present, all junior scientists, are a mix of three men and four women. There is an argument going on about the annual budget. This is conducted mainly by the women. The men interject quite vocally, but no one seems to pick up what they say, and it is two or three of the women who seem to make the final decision. Before the end of the meeting, one of the men looks at his watch and says he has to leave; it is time for him to go pick up his kids from school.

N O SINGLE element in this story is entirely implausible. It is only when we look at them all together that the scenario starts to seem distinctly odd. You will have to visit quite a few businesses to find even the occasional female security guard. Male secretaries are a very small minority in every organization I have ever visited (quite a few!). Female computer scientists are scarce in any business or college, and you will find few of them in top positions anywhere. Where men and women are engaged in collective decision making, the men generally dominate and the women are talked over. Yes, some male managers do take time out for child care responsibilities, but they remain the exception, not the rule.

Let's go a couple of steps further. Imagine a world in which most of the police, soldiers, engineers, air traffic controllers, factory bosses, financiers, boxers, orchestral conductors, and criminal masterminds are women. In this same world the nurses (not their managers), the manicurists, personal counselors, preschool teachers, telephone sales staff, and prostitutes are men.

Perhaps you believe this world is just around the corner and that the gender patterns we see around us are solely the product of cultural conditioning. Yes, culture does have a lot to do with the way things are, and it always will. But culture is a river that follows the contours of our nature. We can dam and irrigate, but only within the limits of the genetic terrain. Only reshaping the landscape by genetic reengineering could lead us into a world of gender neutrality, and even if such a thing were possible, you can bet that neither men nor women would want it to happen.

We could do much more to rebalance the rewards of men and women, but if nursing becomes more highly paid and prestigious than engineering, it is a sure bet that men will start to enter nursing in greater numbers. Men go where the money and status are, even if the work suits them less well than it suits women. Does this mean high pay could turn currently female jobs into male ones? No, many jobs call upon abilities in which men and women differ. Men, for example, are better at the spatial mapping required for air traffic control, and women are more motivated to nurse the sick and elderly. Culture works in tandem with biology.

MALE AND FEMALE— THE HUMAN MODEL

EVERY PRIMATE species has its own mating patterns and division of labor between the sexes. The practice of exogamy, or finding a mate from another kinship group, is a protection against incest, refreshes the gene pool, and builds cooperation between communities by helping to link formerly unrelated kinship groups. Among monkeys and apes, whichever sex inherits the status of the parents, the other sex leaves home. Humans follow the pattern. Like the other apes, we are biologically patriarchal— status passes from fathers to sons. We are female exogamous: it is generally the girls who leave the family home, often departing from their social class, to seek a mate. Even in the most sophisticated Western cultures, the family business along with the family name is generally handed down through the male line, while the daughters pick a different occupation and social sphere, becoming part of a new kinship group upon marriage. There are cultural variations in this pattern to accommodate extreme local conditions—for example, a chronic shortage of either sex or adverse conditions for subsistence—but the general pattern holds.

Function follows form. Another general rule is that the more the two sexes of a species differ in their appearance, the more their roles and behavior differ. In some creatures, the sexes look almost identical and they share their labor with strict equality, like the lonely pairs of albatross faithfully turn-taking to get food and sit on their single egg. Other creatures differ dramatically in size and appearance, like the little male praying mantis who submits to being a postcoital dinner for his big bride (thereby donating himself as a nutritional legacy to his unborn offspring).

Humans fall in the midrange of sexual dimorphism, or visible sex differences. We differ markedly but not extremely in the structure of our bodies. Likewise our minds. This is reflected in how we function socially. To survive and reproduce, men and women do many things equally and cooperatively, but at the same time the sexes typically differ in motives, abilities, and styles of acting. The foundations of our differences reside in our biological systems.

Bodily differences.   We differ in many aspects of physique. Size and strength, distribution of fat and hair, genitalia, and a number of internal organs are all quite different in men and women. Women are equipped for childbirth, nursing infants, and a range of occupations calling for manual dexterity. Men are suited to heavy or athletic activities and to competitive contests.

Genetic differences.   Geneticists are also finding that the genes from mothers and fathers operate differently in how they switch the attributes of male and female offspring on and off. Father's genes bias for fetal size and strength. But big babies are dangerous for mothers, so mother's genes instead switch on the fetus's genes for intellectual and social gifts.

Biochemical differences.   Hormones and mood-affecting chemicals circulate in the bloodstream of men and women in different concentrations, and these change at different rates over our lifetimes. These chemicals have profound effects on the distinctive motives and interests of men and women, and the way they alter with age.

Neurological differences.   Men's and women's brains differ in their structure and operations. Women's brains are smaller and denser, take longer to wear out, differ in patterns of localized functioning, and exhibit greater integration between the two hemispheres. These differences cause men and women to have distinct mental and social preferences and to process information differently.

Sex differences are a problem in organizations not because men and women are creatures from different planets—Mars and Venus—but because we overlap so much in almost all the ways we differ. If we were more different, then we might be able to practice a more complete sexual apartheid, just coming together for occasional mating. On the other hand, if we were identical, we could share every activity equally. We evolved to be alike enough to be interchangeable for many activities and different enough often to choose to do different things or to do the same things in quite different ways.

The overlap between the sexes means three things. One, men and women differ within their own sex in styles and preferences—which is also linked with differences in physique, hormonal balances, and brain structures. Some men are less stereotypically masculine than other men and some women less feminine than other women.

Second, in the middle range there are many occupations that

both men and women are equally happy to do. Jobs in accounting, sales, design, some areas of the sciences, and administration contain no obvious sex bias in ability, though men and women typically bring different attitudes and style to bear upon the same roles.

Third, men and women differ most at the extremes of maleness and femaleness. Computer nerds are almost always men, and preschool teachers are almost always women. The picture is actually a little more complicated than this. On almost any characteristic men differ more than women. In any group the gap between the tallest and the shortest will normally be greater for the men than for the women. The same is true of intelligence and a host of other features. One effect of this is that there are more almost exclusively "male" than "female" occupations and activities. There are more male preschool teachers than there are female computer nerds.

SEXUAL SHOPPING AND MARKETING: THE EVOLUTIONARY AGENDA

A FUNDAMENTAL principle of modern Darwinism is the idea that men and women have a common interest in passing on their genes, but quite different strategies for achieving this goal. Women have a reproductive cycle, a limited store of eggs, a well-defined fertility window of opportunity, and bodies equipped to feed an infant. Men, with their more or less unlimited supply of sperm, have the ability to reproduce over a much longer period. Men and women adapt to these conditions by deploying quite different game plans for securing their genetic futures.

In a world of short or at best uncertain life expectancy, a woman would rarely hope to get into double figures of offspring. Add the natural hazards of infant mortality, and they could expect many fewer than this. In Stone Age times only the most fortunate of women would have two or three of their children achieve maturity. To minimize the hazards to her scarce reproductive resources, a woman needs two things. First are good male genes—i.e., a partner who has inheritable assets such as intelligence, strength, and a healthy constitution. Material assets can also be passed on to offspring, and women may reasonably expect men who accumulate them to have done so at least partly because of their "good" genes. Second, the chances of a woman's children surviving will be greatly aided by her being able to count on reliable help in rearing them, especially during the first few years of the child's life. So the woman is also motivated to look for a partner who is loyal, cooperative, and willing to provide for them. In short, women need to be choosy about their men.

In fact, women's choosiness extends to all relationships, since it is through their networks that they acquire reliable intelligence about people and resources. Women therefore need sharp intuitions about character—which means the skills of empathy and communication. They need these to check out the truth or falsehood of men's signals about their qualities. Not for nothing do women spend more time than men testing reliability and trustworthiness before committing to relationships.

From a woman's perspective it also makes sense to have a wide range of options. This increases the motivation for women to be good at navigating social networks. They are also motivated to attract men, so as to maximize the number of possibilities available to them and ensure that their precious reproductive resources are not wasted. To this end, women need to send signals that will attract the right sort of man. Some of these are physical, advertising their fitness for reproduction. Research has shown that men judge the attractiveness of women on a common set of features across all cultures and times: body symmetry and other cues that suggest health and fertility. Unconsciously women know the formula and decorate and display themselves to enhance these features. Women also behave in ways that are likely to appeal to men. One such strategy is doing things that make men feel special, strong, and admired. Women are thus tuned in to the somewhat different game men are playing.

Men try to be competent, confident, and strong. Because their reproductive resources are less limited, they can afford to be less discriminating. Quality also matters to men, but, unlike women, men are inclined to maximize the number of opportunities available to them. Like women, they look for signs of good genes: physical attractiveness, which signals health and fertility. Men also place a high premium on fidelity, a partner devoted to them and their children. Knowing that women, especially those of the highest quality, are disposed to be choosy puts men in competition with one another. Men who are the most conspicuous achievers do best in both quality and quantity. This raises the stakes for men; they are driven by the need to do as well as they can in status, achievement, strength, and wealth. This may mean taking risks.

FRANCES'S TALE: AN EVERYDAY OFFICE STORY

Frances is an information officer in a public service agency. She advises people who drop in or make phone inquiries about welfare rights and legal and health issues. She is a quiet, conscientious person in her mid-forties who has worked for the agency for around twenty years. She works alongside Jake, who is about ten years her junior and who joined the agency three years ago. Both are supervised by Betty, who runs the office. Frances and Betty are friends; they are the same age and joined the agency at the same time. Betty is a rather anxious woman who bustles around with great energy. She spends little time in the office and is frequently away at meetings with central administration or visiting other offices. She relies heavily on Jake and Frances. She often seems distracted and slightly overwhelmed by the demands on her. Actually, Betty relies more on Frances than on Jake. Frances is a calm, even-tempered woman—hardworking and positive, seldom fazed by stressful situations. She is intelligent, practical, and well organized. She basically runs the office and keeps people happy. She is universally liked. She is not ambitious but does want to be respected for what she does. Her family is her main priority. Jake is ambitious and a lot less conscientious than Frances. He likes to play with the computer systems and figure out new ways of organizing databases. He is quite uninterested in the clients but can turn on the charm when and if it's needed. He doesn't attempt to conceal from Frances how he falsifies office records to make it look as if they have been busier than they have. He flirts with Betty and regularly asks for feedback sessions with her. Much of the time Frances finds herself covering for things that Jake has missed.

It is the time of the annual appraisal. This interview ends, as it usually does, with Betty's saying words to the effect of "keep up the good work." A couple of days later Frances is filing some papers in Betty's office and she accidentally sees Jake's appraisal form, which Betty should have filed away herself. Frances can't help noticing the rating at the bottom of the form: "outstanding contribution." She reads no further. It is not her nature to pry, and she believes in strictly professional conduct about confidential information. But she has seen it and she feels aggrieved and undervalued.

What should she do? Her instinctive response is clear: do nothing. This is no business of hers. Resignedly she accepts that it is better to carry on with your job and live your life without creating conflicts. One of her feminist friends tells her she is an oppressed woman and that Betty is, too. Frances disagrees. For her, it is a matter of personal integrity. She can't see how she can change the world by getting into a fight over Jake's appraisal. Besides, for all his faults she quite likes him and wishes him no ill. She has her own priorities, which have nothing to do with work. Let Jake get what he wants, she feels, and good luck to him.

The story is unexceptional. Of course, it is a reflection of a particular office and the specific personalities within it. Such incidents always are. Yet a similar drama is being acted out in organizations the world over. A thought experiment helps to show this. If we turn Betty and Frances into Bert and Frank, and Jake into Jane, how much more implausible does the story become? Would a Jane go to the same lengths as a Jake to get ahead? Would a Frank be as inclined as a Frances to opt for peace and quiet?

How much of this is just culture? Or are we seeing here, as elsewhere, how culture is shaped by psychological bias? Here this takes the form of the stereotyping of older women, female preference for harmony over competition, and male inclination to risk-taking and deviance to advance their status.

THE INVISIBLE GRAVITY
OF INSTINCT

EVOLUTION IS blind to the future. Adaptations are always retrospective. Something that worked in the past is retained as part of the repertoire of design features until a better feature comes along. Evolution stocks our minds not with forward-looking strategies but with desires and feelings which lead us into situations in the past that turned out to be useful, and steers us away from others that have brought frustration and loss. The motives we feel consciously are such things as sexual attraction, the urge to win in competition, the desire to be liked and trusted, and so on.

We have no mental modules for our genetic interests. Even if we correctly identify our motives, this is as far as we can see; we have no need to be aware of the evolutionary goals that shape our motives. Rather, they operate through our psychological makeup like an invisible gravitational force. Our instincts are pretuned to those behaviors that bring about good reproductive outcomes: for men, winning status in competition with other men; for women, having good networks and the emotional radar to enhance choice.

The Darwinian theory of sexual selection shows why design features to survive and reproduce are not enough. Your genes have to win out in competition with those of others of your sex for the best mating opportunities. In a world where lots of men look as if they could be good investments, how does a woman choose the best? And where lots of women look good, how do men pick which to court? Sexual selection sets off an evolutionary arms race leading to the possession of costly attributes, such as the peacock's tail, to advertise quality. Men and women (especially men) are driven to ever more extravagant displays to signal their worth and attractiveness.

How does a man signal his worth? By high levels of achievement. Men believe that they have to be better than all the others on show: they must run faster, work harder, spend more money—even to the point of wrecking their health and well-being. Men the world over are driving themselves to early graves in pursuit of wealth or reputation, compulsively leading lives that leave them no time to consume the benefits they create. Women, for their part, cultivate the qualities that will give them an advantage over others long after there is any reproductive need to: paying elaborate attention to their appearance and dress, building social networks, and finding opportunities to conspicuously demonstrate their moral worth.

In later life these drives do mellow, but not by much. Evolution works to get design features in place so they can yield a benefit, no matter if they become redundant after they have served their purpose. It would cost too much in extra design engineering to have them switch off, especially when people don't live much past their expiration date. Hence we get the spectacle of old men posturing like young bucks and women continuing to act girlish well past their youth.

SEX DIFFERENCES AT WORK

IN THE world of hunter-gatherers, men can be only intermittent cocarers for their offspring. In all hunting and gathering societies described by anthropologists, big-game hunting is exclusively done by the men, much of it far from camp. Child care restricts women to activities closer to home; hence women concentrate on foraging in their immediate environs. Hunting and foraging require distinctive perceptual motor skills—mental mapping for hunting (men), finding and remembering the location of objects for foraging (women). Tests confirm that men and women differ markedly in these skills, strongly suggesting they are hardwired.

Hardwired is a shorthand term, at best a metaphor. The instincts and gifts we possess are potentialities, not circuit boards. The experiences of a lifetime, especially those of childhood, trigger the distinctive awakenings of human nature. Evolution shaped our potentialities. Much of our learning is automatic—no special instruction is needed. We absorb what we need to. It has been noted how much easier it is to teach a child to fear snakes than to fear automobiles—learning to fear another creature maps onto a preset bias. Experience sketches our awareness not on a blank slate but on a prepared surface.

If the sex differences we have been discussing are hardwired, we should expect to find evidence for them across cultures and over lifetimes. So there is, in study after study. Take, for example, the work of linguist Deborah Tannen, who has explored in detail how children talk to other children and adults to other adults. This exposes fundamental differences in the way boys and girls deal with each other: boys focusing on competitive games and projects, sharing experience through action; girls playing empathy games in which the sharing of feelings predominates. Tannen explicitly disclaims an evolutionary perspective for her work, but there is a close fit of her data with what the new Darwinism has to say.

Here's a summary what we know about the differences between the sexes:

List A

◇ Men are bigger and stronger than women.

◇ Men don't live as long as women, and they lose their mental faculties earlier.

◇ Men are more promiscuous than women.

◇ Men are more competitive than women.

◇ Men are more self-interested than women.

◇ Men are more likely to take risks than women.

◇ Men like to play games more than women.

◇ Men have more skill and interest in using heavy tools and weapons than women.

◇ Men focus on tasks in a more single-minded fashion than women.

◇ Men have better mental mapping skills than women.

List B

◇ Women feel and express emotions more easily than men.

◇ Women are more empathetic than men.

◇ Women are more dextrous than men.

◇ Women are more nurturing than men.

◇ Women have better social skills than men.

◇ Women have better verbal skills than men.

◇ Women are more cooperative than men.

◇ Women are better at multitasking than men.

◇ Women have a better memory for object location than men.

◇ Women are better at spotting embedded objects.

But what about the women who score higher on the A list than some men, and all those men who score higher than some women on List B?

True, these are not absolute differences. Lots of men have good social skills. Many women are skilled map-readers. But even slight average differences in a population produce exaggerated effects over time. We tend to sort into groups of clustered gifts and preferences, even when these differ only slightly in magnitude. Although there are many women who are taller than many men, almost all sports where height could influence performance are strictly segregated (basketball and volleyball, for example). The same applies to psychological differences. You can take just about any of the items on the A and B lists and find social roles sex-segregated on the basis of them. Note that this does not always result in a favorable allocation of roles to men. All the most risky and dangerous jobs in the world are done by men. Enough men are attracted to roles where they can exert themselves physically or feel heroic to be the first to volunteer to be the firefighters, lifeguards, and commandos of society. Men also do more of the worst-paying and most miserable jobs. Because men are more widely spread than women in abilities, not only do men predominate at many of the pinnacles of high achievement, they are also much more populous in the very lowest occupations.

Both men and women have instincts to nurture, but women nurture more strongly than men. Both women and men have mapping skills, but men more than women. But the result is an almost exclusive assortment of women into caring professions and men into occupations requiring spatial mapping. A startling 93.5 percent of the nurses in the United States are women, while 97% of air traffic controllers are men.

The differences between men and women have big effects, but not to the point of dividing us into groups with exclusive rights over our psychological territories. That would be a stupid evolutionary outcome for the cooperative life we have to lead.

Indeed, it is the need to cooperate with a flexible division of labor in all the tasks of the clan that exploits the differences *within* the sexes. The differences between men and between women give us an extended range of options about how to organize our relationships.

We discovered some thousands of years ago that by organizing in certain ways we can do well economically—that is to say, produce and consume more things. We are prepared to pay a high price for this in terms of doing without ease and fulfillment. But men and women differ in their willingness to trade quality of life for economic benefits to themselves and their families.

> In 1999, *The New York Times* reported how a woman colonel in the U.S. military unexpectedly quit just when she seemed on the brink of becoming the army's first woman general. The stress of working twelve-hour days through a series of fast-tracking overseas postings had eventually defeated her. She feared that she was becoming too removed from the life of her teenage daughters and that they were withdrawing from her. She said, "I found myself saying I would be a better commander if I wasn't a parent and a better parent if I wasn't a commander." The demands of army life in this case took no account of her status as a mother. These included postings far away from home on short notice, daily physical training, and being on call all the time. "Men treat you differently when you're pregnant, asking you whether you can really do those push-ups and whether you can take that trip."

The military life might seem to be an extreme example, but many businesses make few allowances for women, and those who want to succeed are reluctant to press for concessions for fear of being stigmatized as taking advantage of their gender.

As we have seen, the organizational model is male, in the sense that it is based upon features that predominate more in male than in female psychology: technical focus, single-mindedness, competitiveness, a desire for control and hierarchy.

Because of this, in many areas of business women do not stand a chance of fair and equal treatment. Their plight is not helped by the fact that many of the men around them feel frustrated and undervalued.

Instead of seeking the uniformity of men and women—an unrealistic goal, because we persist in wanting and liking different things—we should focus on equality of opportunity and freedom from oppression. This means recognizing rather than denying sex differences in how we think, feel, and act.

THE GENDER BALANCE EFFECT

THE DIFFERENCES between the sexes are most obvious when men and women are segregated. When men and women are together, both sexes are inclined to engage in sex-typed display (dressing up, boasting, flirting), but other elements of their styles of behavior are more moderated by each other's presence. This is clearly a bit of a strain. It is not long before men drift into men's talk and women into women's talk. Most of us have had the experience of attending a dinner party where the women congregate at one end of the table and catch up on what's going on in their lives, while the men go into a huddle and talk about sports or their investments.

The maleness and femaleness of behavior in sex-segregated groups are visible in many businesses. To survive as a female police officer, soldier, or engineer requires an exceptionally thick skin. On a TV talk show about racism in police forces, a black male officer was encouraged by a sympathetic black interviewer to recount the hard times he had had. He thought for a while, then said, "Well, you get called a few racist names. It's not nearly as bad as what the women have to put up with." Race differences—whatever trouble they cause—are skin deep. Gender differences are fundamental.

A similarly tough environment exists on Wall Street. As one of only two women out of fifty on the bond-trading floor of a leading investment bank, Susan Estes found herself being ogled and the butt of her male colleagues' coarse jokes. She hit back by shaming them, recording and publicly recounting their comments. This was the first step in transforming their attitudes as well as their behavior.

Any man will tell you that the way men get along together—joshing one another, arguing, doing deals, telling stories—has a boys' club character to it. To see this, join any all-male gang of builders, salesmen, lawyers, or medics on a night off. If even one woman joins the company, the character of the party changes immediately. Some men—those who feel uncomfortable with the male locker room culture—breathe a sigh of relief at this. Women have told me that the reverse happens and they find it infuriating. A group of women are all behaving sensibly and cooperatively until a man joins the group; then a certain number of them start acting deferential, flirtatious, or strident in the pursuit of male attention. Generally, though, because of their superior social skills, women find it easier to operate effectively in mixed-sex groups than many men do.

You can see marked differences in the ways all-male and all-female groups prefer to work. Take a bunch of executives or business students and ask them in single-sex groups to choose, plan, and implement a task. What the men are likely to do is take turns in suggesting and countersuggesting strategic approaches. One individual will probably take a leading role in arranging a compromise between the leading contenders. Then they will divide the work, assigning roles and giving practical support so that cooperation is highly focused and intermittent. There will be a good deal of mildly aggressive joking, a kind of rough playfulness in which sharp exchanges seem quickly forgotten.

The women, in contrast, will spend much more time talking about what each of them thinks, listening and expressing support for one another. Someone who disagrees with the strategy opted for is less likely to be simply overridden than in a male group. She will be persuaded or accommodated. In doing the task, they are likely to cooperate more continuously and with less role segregation. The feeling of the group will be warmer than in the male group, but if there is conflict, any wound will be deeper and recovery from it will take longer.

The men will be more efficient, in terms of the time taken, as long as they don't get locked into conflict. The women will reach a better solution as long as they don't try to accommodate too many contrary points of view in some kind of a messy compromise. Put men and women together in something like equal numbers and you lose the elements of gender polarization. Men and women temper each other. If they are receptive, women can learn from men the benefits of suspending the expression of emotion for the sake of getting things done. Men can learn from women how much more effective they can be by tuning in to how other people feel before jumping in with both feet. Offices with roughly equal numbers of men and women are healthier. They provide opportunities for mutual learning and adjustment.

Let's look at how sex differences in business reveal themselves most dramatically. It is on the two axes of organization: the vertical axis of status and the horizontal axis of function—who does what.

HIERARCHY AND STATUS

FROM TIME to time newspaper articles tell us about how women are breaking through the glass ceiling. But these instances continue to be treated as newsworthy *because* they are rare. Hard evidence for significant change remains elusive. In the United States and elsewhere, record numbers of women are finding their way into the middle and upper echelons of management, but not to the topmost levels. The number of female chief executives in business remains around 5 percent, and this woefully small figure has remained consistent for a long time. In 1999, 11 percent of the Fortune 500 board members were women. Only a tiny fraction of these were chief executives, and only one of these was in a major blue-chip corporation—Carly Fiorina of Hewlett-Packard.

A weekly feature in the *Guardian,* a national U.K. newspaper is an interview of a CEO who responds to a standard set of questions. These include such items as "What was your biggest mistake in business?" "Who is your management guru?" "Is there a glass ceiling in your organization?" The answers to this last question are revealing. In just about every case the CEO gives a firm negative answer. Surprise, surprise—almost without exception the respondents are male. What is going on here? Are they deceiving us or themselves, or are they right? How long will it be before a woman comes to occupy the position they currently occupy? If we repeat the survey in ten or even twenty years, will the respondents be at least 50 percent female? Would you bet money on it?

So who is kidding whom? I would like to take at face value what these men are saying—that they sincerely care about the advancement of women in their organizations and that they have done everything they can to ensure that no barriers to female advancement exist. A typical response to the questionnaire was "No. Anyone can have the top jobs. They just have to be good enough." Okay, but good enough at what, exactly? One key attribute is wanting the top job enough to be prepared to do all the things that might deliver it. Most companies stop short of affirmative action but sincerely try to maintain gender-neutral methods of executive selection, development, and promotion. Still the women fail to come through. Why?

First, in the classic hierarchy, status is a zero-sum contest— a game where one player winning means another has to lose.

Women in competition with men for these positions are likely to come across men ready to fight dirty to get ahead. Even fighting cleanly, men have the buddy advantage of knowing their networks, and being more surefooted in playing snakes and ladders in the male-dominated corporate hierarchy.

Then there are the material barriers for any ambitious woman. Again, culture allies with biology to conspire against them. Bearing and rearing children cost women time when they could be building their résumés. Then even in the most enlightened subcultures there are unequal expectations about how men and women should respond to the demands of family life, including how husbands and wives should accommodate and support each other's careers. Au pairs and child care centers do not preclude women's time and energy from being claimed by school holidays and family illness. Women then find that the attention they pay to personal concerns makes men, and even some unencumbered female colleagues, view them as less than 100 percent for the company. The phenomenon of overwork exposes the difference even more. Family men are freer than family women to attend late-night meetings, weekend retreats, and the like. Male psychology makes it easier for men to remain focused on a goal.

Thus the numbers of women available for selection to top positions are winnowed down. What is left is a small population of survivors—women who don't have domestic responsibilities but do have psychological profiles that prepare them to think and act like men in order to move up. Many "career women" will confess to you that they behave this way with deep inner misgivings about what it is doing to them and how it is unbalancing their lives, as in the case of the female colonel in the army that we looked at earlier.

Indeed, motivation is the biggest barrier to women's advancement. Even women poised to reach the top often stop and ask themselves if this is what they really want. They see what it takes to climb the last few rungs of the ladder and wonder if they really are prepared to do so. And even if they win, do they really want to play power politics to hold the top slot, working long hours with little or no family life?

So what is a woman to do who wants to build something she can believe in? How can she assume a leadership role in an organization and not feel like an alien? For many women, the answer is to quit and start their own business or to find the company of others trying to create alternatives to the typical male organization. This is just what legions of female executives are doing— leaving large corporations and founding their own more women-friendly businesses. There has been a steady trend in North America and Europe for female business foundings to outstrip males' since the mid-seventies, when women began to enter management in large numbers. Many professionally qualified women are also joining consultancies and service businesses that have broken free of the classic model. Many feel the environment allows them to be themselves, influential and fulfilled rather than the token member of the boys' club.

CHOICE OF ROLES AND FUNCTIONS

HOW MUCH is the gender typing of roles in society the result of brainwashing? Does it come from the toys we give our children, the stories we read them, and the images they watch? Do the expectations of parents and teachers decisively push boys and girls in opposite directions?

There are several things that don't add up here.

One is that this is an inaccurate account of what happens to children. People actually bring their children up in lots of different ways, and the results are at best unpredictable. Moms and dads devoted to the ideal of gender-neutral parenting have time and again been disappointed with the results of their attempts. They give Rose a construction kit, and she makes a doll's house with it. They give Roy a doll, and he turns it into an action figure. Boy babies are different from girl babies before they can speak, watch television, or be reasoned with.

A second problem is that the brainwashing theory depends on the notion that a human being is a passive vessel. Rather, we are programmed to be the agents of our own development, and the process starts decisively in infancy. In fact, parents are conditioned more by their children than vice versa! After all, Mom and Dad are eager to keep Baby happy long before Baby can figure out she can keep Mom and Dad happy. From when we are first aware of the world, our unique temperamental biases lead us to choose and keep choosing our own paths—selecting the friends we like hanging out with, playing the games we enjoy, practicing the skills we are good at.

The third problem is that if adults have gender expectations, where do these come from? Are they brainwashed, too? Where does all this brainwashing come from? Actually, parents do treat boy and girl babies differently. Much of this is quite unconscious. Mothers have been found to talk differently to male and female babies without being aware they are doing so. We cannot help but provide different gender models for our children to observe and imitate. Even when we are conscious of our gender biases in how we treat each other, it makes much more sense to assume that we are responding to some inner knowledge, not that we are reading a script handed down to us.

In earlier times the social roles men and women were allowed to perform were severely restricted. Now most of these restrictions have been lifted; boys and girls and men and women can choose to do what they like and are good at. But these differ consistently, as we can see all around us. Language and humanities courses are full of female students. Engineering and technical courses are packed with males. Nursing and teaching are populated mainly by women, production and computing by men. Women gravitate toward advertising, retail, communications, and

the helping professions, men toward manufacturing, heavy industry, and applied science.

Within organizations, where are the women? Wherever
interpersonal skills and relationships are at a premium. So we
find women concentrated in human resources and public relations, and also in many customer-facing roles, which just happen
to be located at the bottom of the hierarchy. Men are in the technical command and control roles, which also just happen to lead
to the executive positions in most companies. This is a sexual
division of labor that can be found almost universally in the business world, including those firms that most assiduously promote
equal opportunity. No amount of legislation to advance the
careers of women can counteract the choices that men and
women themselves make and what they feel most comfortable
doing.

The separation is pronounced but not complete. Men and
women make equally valuable but distinctive contributions in
many areas of business, such as finance. Women and men have
equal facility with numbers, but women have the edge when it
comes to remembering, organizing, and working with detailed
information, and they are generally better than men at managing
relationships with clients. Men, on the other hand, are more often
drawn to areas of strategic control, risk taking, deal making, and
competitive decision making. Many clever and highly motivated
women find their way into the fast-paced world of investment
banking, but you will find them almost exclusively in relationship
roles—research and sales. The aggressive high-wire business of
trading is overwhelmingly male.

In the words of one trading floor manager, "There are times
in market making where you have to go after and kill some rival
out there who's trying to do the same thing to you. It's a great
feeling when you succeed—when you know you've got him by
the balls. Quite honestly, not many women want to get into that."
But tellingly, this does not make men better traders, only tougher

ones. In fact, there is evidence that men overtrade more than women in financial markets. Their drive to achieve leads them to waste more time and effort, and this has an efficiency cost. So why are there not more women traders? They might be better at some of the fundamentals, but a successful trader has to be able to stand the incidentals.

DOMINATION AND LEADERSHIP— FOR MEN ONLY?

THE ANGER of many feminists is entirely justifiable. In many workplaces men get ahead by doing things that women would find repugnant. Elsewhere, sexual politics lead men to demand and women to offer a mixture of subservience and psychological massage. The secretary-boss relationship still embodies this ethos in firms all around the globe. More extreme manifestations are reported daily in cases of bullying and sexual harassment. These are almost without exception one way, from men to women, and are most likely to occur in male-dominated occupations. Men give free rein to their impulses for power and domination in such organizations.

Women are also interested in status and are quite capable of playing power games, but they do so in ways that are typically different from men's. They engage less in direct contest and work more through the politics of relationships and shared values. It is quite possible, and not uncommon in family life, for women to use their gifts to dominate men. The model of women as mothers or matrons is based on the ego strength, social intelligence, and organizational gifts that many women have in abundance. However, the rules, practices, and structures of business organizations favor males and their aspirations.

The way most organizations choose leaders screens out women or makes their way harder at each successive level of the hierarchy. Sometimes women do make it through to the top of the most progressive corporations. Carly Fiorina did it in Hewlett-Packard. Charlotte Beers did it in the advertising firm Ogilvy and Mather, paving the way for a female successor in Shelly Lazarus. These are isolated exceptions, though. A woman who really wants to be a leader has two alternatives. One is to make personal sacrifices and compete like a man to move up. The other is to quit the corporation, branch out on her own, and establish her own model.

When women do the latter, the results are often impressive. Two high-profile women from either side of the Atlantic—both running highly successful cosmetics businesses—are Anita Roddick, who heads the Body Shop in the United Kingdom, and Mary Kay Ash, who runs a direct sales business in the United States. Both offer a nurturant model, reaching out directly to people as guardians of morals and faith, albeit of different kinds. Ash's business is founded on her strong Christian values, Roddick's on a passion for environmental causes. Both nurture their people. In 1991, Roddick said the following about her business philosophy:

> I am mystified by the fact that the business world is apparently proud to be seen as hard and uncaring and detached from human values. . . . The word "love" was as threatening in business as talking about a loss on the balance sheet. Perhaps that is why using words like "love" and "care" is so difficult in today's extraordinarily macho business world. . . . I think all business practices would improve immeasurably if they were guided by "feminine" principles— qualities like love and care and intuition.

Through the power of their personality and personal vision, these women built their businesses by attracting to themselves people who shared their vision and could relate to their values.

Mary Kay Ash founded her direct sales cosmetics business after a series of personal misfortunes: her husband left her, she was broke, and she was passed over for promotion by her employer. Further emotional trials and separations followed. Her Christian faith became a sustaining force in her life and business. A critical starting point was the image of her mother as a nurturing role model. Ash was also moved by her empathy with the working-class women of small-town America, whom she recruited for her business. These were women whose lives had denied them the opportunity for financial independence and occupational satisfaction. Ash created a business model that resonated deeply with these women. The Mary Kay organization offered them a psychological lifeline. To make this happen required the charisma of an exceptional leader, whose strength of character had been forged by a life of adversity.

These examples apart, women leaders remain extremely scarce. Domination, competition, and patriarchy are biologically encoded as our model of authority. These values are integral to the promotion systems that operate in business and public life. Wherever we create an opportunity for a single leader of a hierarchy, nine times out of ten it will be men who strive for, attain, and hang on to the role. But good leaders are scarce. As the demand for collaborative, communicative, and interpersonally skilled leaders increases, so should the numbers of women taking up top positions. But for this to happen, big changes will be needed to avoid talented women being screened out or excluding themselves just when they might be needed most.

A NEW GENDER ORDER IN BUSINESS?

WHAT DOES our analysis bode for the future of organizations? There are many positive indications for women in business. More and more women are entering the labor market.

Education is becoming less gender-typed in shaping the choices of men and women. The most overt forms of discrimination in selection and promotion are being avoided. The law has also helped. More cases of male violence and sexual misconduct are being detected and deterred. The huge growth in knowledge work is providing more arenas for men and women to work alongside each other as equals. Across the business world there is a greatly increased emphasis on relationships—dealing with customers and clients, building teams, and forging alliances.

The complexity and uncertainty of today's increasingly competitive global markets, in which information technology has also lowered the barriers to entry for new firms, are placing a new premium on a strong and flexible business culture. Companies are realizing that the key competencies they need are no longer purely technical or based simply upon people's authority to command. Interpersonal and communication skills are now the key to competitive advantage in many businesses. These factors mean that there has never been a better time to be a woman in the business world. So we should really be able to look forward to a new organizational order in which women play a significantly greater role.

Yet our analysis suggests some things will not change. The sexual division of labor will continue to be found because women and men prefer different kinds of roles. Leadership will remain a male preserve in many organizations. Although we are seeing increasing numbers of women in the middle levels of most businesses, as long as the status hierarchies persist, men will strive hardest to reach the top. Who gets there depends on the interplay of three things: what people want their leaders to do, what kind of human raw material is available for leadership positions, and how leaders get chosen. It is time to look at these elements more closely.

◆ 4 ◆

The Natural Selection

of Leadership

HUNT THE LEADER

THERE IS a big disconnect between what pundits say about leadership and what people see going on around them in business. Management consultants, writers, and educators perpetuate the myth that by applying the right knowledge, training, and resources, any manager can be turned into an effective leader. Yet we seem to live in a different reality. Pause for a moment and think about managers you have worked with. Is it not striking how great the differences between them are? On one side are talented people with drive and skills, people who look as if they would be successful wherever they went. On the other side are assorted solid citizens who might be effective bosses under certain circumstances, and a third group who you reckon never could or should be leaders, so deficient are their temperament, motives, or skills. It is difficult to conceive of the training that could possibly transform the third group into the first.

At the same time, without being unduly cynical, you may reflect that there isn't always much connection between people's talents and the roles they actually occupy. We have entered the realm of the Dilbert cartoons, where "the most ineffective workers are systematically moved to the place where they can do least damage—management." The reason this strip became one of the most popular management commentaries of the 1990s was that it gave the lie to the gap between management theory and practice, especially in the realm of leadership. It resonated with all of us who have endured bosses totally lacking in leadership qualities, witnessed the good and talented being passed over for political reasons, or seen people with vitality and ambition crushed by corporate bureaucracy.

But let's not get ahead of ourselves. First let's define just what a leader is. The literature on the subject is confusing because of the many abstract ways in which leadership is defined. Such clichéd abstractions as "doing the right thing" or "making a difference" aren't a lot of help. So let us simply define leadership roles as the positions of highest authority within a social group. This objective definition points to leaders in many quarters of every society and organization. Here we shall be talking mostly about the person who has the highest authority within a largely autonomous group—for a tribe the chieftain; for a country the head of state; for a corporation the president. Within a large corporation, heads of divisions and leaders of some self-governing teams also fall within the definition.

THE UNIVERSAL LEADER

WE HAVE been let down so many times by evil or ineffective leaders. One is impelled to ask, can't we do without them? However disillusioned we may be, the answer is that we demand leaders. Dominance is a biological universal among all social mammals. It takes different forms, from chimpanzees who beat up on each other for alpha status to wolves who try to outconciliate one another to gain the regard of the pack. In human society we have such extremes—a wide but not limitless range of leadership types. In business we try, with some success, to design organizations to be leader-proof but never leaderless. Create a society without leaders, and they will just reappear in all but name.

Here is why we need leaders and the reason we so often get leaders who are incompetent or domineering.

The Biological Lure of Status

Hierarchies are common throughout the animal kingdom and universal among primates. Great benefits of health, resources, and well-being accrue to those of high status, and severe penalties and stresses afflict those of low status. This is the lure. The more complex the social system, the more routes there are to high status. Humans can gain high status by inheritance, by possession of a talent or skill, by alliance, and by social dominance. Because in the world of work many of our hierarchies are monolithic—a single chain of command from the boss down—the routes to high status get bundled together. This bundling together of pathways means that you can get to the top by being fit on one strand and unfit on others. This is why the heir to a family business can turn out to be someone with no talent for or motivation to lead. It is what induces people who want the benefits of status and recognition to take on the burdens of leadership, even when it doesn't suit them. The monolithic nature of our hierarchies is what gives us all the lousy leaders of business empires who started out as brilliant engineers, financiers, and computer wizards.

Evolutionary Archetypes

The model of the leader is deeply embedded in our psyches. It comes from our experience as children looking up to powerful adults on whom we depended. This image remains with us into adulthood—the vision of someone who will protect us, solve our problems, bring us what we desire. Magic, religion, and romance all draw from this psychic well. We search for heroic leadership time and time again—so much so that we project all kinds of mythical attributes onto quite ordinary men and women in leadership roles. Some leaders are not far off the slow-witted gardener in the movie and novel *Being There*, who arrives at the brink of the U.S. presidency on the strength of the vacuous horticultural metaphors he utters. His simple truisms are eagerly consumed by a group of political hacks desperate to fill a power gap, and by a wider public yearning for wise leadership. It seems as if we will tolerate leaders with quite limited competence. Most monarchs have been revered regardless of their personal qualities—as long as they don't behave inappropriately, such as by being too familiar or appearing weak.

Dominance Is Hardwired

The vocabulary of dominance is universal—an upright stance, decisive gestures, strong voice, and an unflinching gaze. Each species has its own repertoire. We have ours and recognize it instinctively. Men, especially, have years of practice in demonstrating and responding to dominance and quickly learn the limits of their own powers. Dominance extends beyond body language and physical contest. It also comes through in how we communicate with and relate to one another. Dominant people talk more than the others in any group—what has been called the babble factor.

People who would be leaders try to master the attributes of dominance. For some this comes easily. For others it is an impossibility. But the chemistry between dominant leaders and followers, the love affair called "charisma," doesn't only give you Nelson Mandela, it also produces Adolf Hitler. Dominance is a hairsbreadth from domineering.

Leaders are also quick to realize that their dominance cannot always be displayed in face-to-face interactions. It requires symbolic proof.

> Chris Argyris, an eminent elder statesman of organizational psychology, told me of the time he met Jimmy Hoffa, the notorious leader of the Teamsters' Union in the 1950s, in his palatial Washington, D.C. office. Chris remarked to Hoffa how pleased he was to meet him and to discuss issues of mutual interest, especially related to the labor movement. Hoffa knew that Chris had just been talking to Tom Watson, the head of IBM. Chris was startled to find Hoffa turning the conversation immediately to the one and only thing he wanted to know: "Tell me," asked Hoffa, "is Watson's office as big as mine? I bet he hasn't got such a fine view and such good furnishings." He was right, said Chris. Watson hadn't. The shock for Chris was to find that this was all Hoffa cared about.

Hoffa was right in another way. Like many leaders before him, he felt that his trappings were the visible symbol of his worth. The pride of citizens is embodied in the leader. It diminishes *them* for their leader not to look the part, for they share in the greater glory of the leader's wealth. This is the real reason big companies pay their top executives inflated salaries—not as an incentive for leaders to perform but as an indication of the worth of the company ship and all who sail in her. It is corporate flag-waving. Like the peacock's tail in Darwinian sexual selection, the fitness of the company is demonstrated by the magnificence of its leader's adornments—in this case the leader's bonuses.

In my early days as a researcher, I asked some militant trade union members what they felt about the owner of their firm's driving around in a very ostentatious white Rolls-Royce. In my naïveté, I was taken aback by their response: "Good luck to him. He can spend his money as he likes. Besides, it makes us proud to see the company doing well. We just want our share." Such displays plug into our need to feel secure and attached to power.

THE ESSENCE OF LEADERSHIP

Once upon a time, on a tropical island, there were two young men studying in the same academy. Both were from well-to-do families. One had natural brilliance, inborn athleticism, good looks, and great powers of oratory and persuasion. The other was more anonymous—clever, quiet, hardworking, determined, and independent-minded. The first, a natural and charismatic leader, went to the local university before entering politics. The second went to America to study chemical engineering, returning to take a junior technical management job in the local office of a large U.S. company, turning down the opportunity to be groomed to take over the family business. Both turned into great leaders who changed the world in different ways. The place was Cuba. The time was the early 1950s. The two men were Fidel Castro and Roberto Goizueta. The one transformed his country and significantly influenced the politics of the late twentieth century. The other, over a sixteen-year period as CEO, transformed the fortunes of the best-known company in the world, Coca-Cola, growing it from a $4 billion to an $18 billion business.

So what happened to Goizueta? What kind of leader was he? Did he undergo a transformation to achieve this, or was his leadership talent always there, latent and invisible? Was it also inevitable that Castro become a leader?

To answer these and other questions, let us think about what the core elements for leadership are. Three matter: drive, ability, and constitution.

Drive

Leaders are driven, but not by the same things. For some it is the drive to have dominance over others that comes easily and naturally to them, as it did for Castro. Others are driven by the desire to acquire status by distinction. These are more impersonal needs to achieve, usually via demonstrated excellence in some particular field. These two drives, dominance and status, are quite different, but both lead to the top. The first gives power over people. The second gives reputation and eminence. Put them together, and you get a chart like the one below.

Political leaders such as Castro and many business leaders—for example, Lee Iaccoca. Sam Walton, Jack Welch, Bill Gates, and Steve Jobs—fall into Type A, *dominant bosses*. They have the desire to dominate plus the drive to achieve through competitive striving. Was Goizueta one of these? Partially. He displayed a single-minded determination to succeed, and when he got the chance to wield influence over people and take responsibility as their leader, he seized it.

Four Leadership Types

Type	Dominance Drive	Status Drive
A: Dominant boss	high	high
B: Ambitious professional	low	high
C: Informal influencer	high	low
D: Reluctant leader	low	low

Some leaders, though, remain stuck in Type B, *ambitious professionals*. They have little interest in dominance over others. They are leaders chiefly because they possess some technical gift. These are the faceless technocrats you find running some investment banks and science-based firms. These gifts can be quite colorful, like the creative people who get to head up media firms and arts organizations, but really have no interest in managing people. Also in this category are the tough achievers, those who like to compete and win but have little interest in leading. We might call this the Coriolanus syndrome, after the great warrior of Shakespeare's play—a chieftain of great pride and honor who refuses to do the PR work necessary to keep his support high among a clamoring populace. They interpret his attitude as one of disdain and bring him low for it.

Type Bs only survive for any length of time in a business when the need for a dominant leader of people has been set aside. This can occur when a business runs like a machine through its operating rules and procedures—what has been called the "machine bureaucracy" model. It can also happen in dynastic businesses where everyone is content for the top person to be more a symbolic than a real boss, while people at lower levels are busy pulling the strings. Type Bs are also a safe stopgap after a company recovers from a bruising spell under a brutalist Type A leader. In business, as in politics, charismatic leaders are often followed by peacemakers, such as happened when Sunbeam appointed a quiet financier to succeed as leader following the firestorm of Al "Chainsaw" Dunlap's tenure.

Type Cs, *informal influencers,* usually don't rise to leadership positions, mainly because they don't want to compete. They are not driven by challenge, often because they fear the risk of failure. In the battle for dominance or status, the higher you get, the more exposed you are and the harder you can fall. But the dominance drive of Type Cs makes them potentially dangerous people. You can find them in every organization. They like having power without responsibility. Sometimes they are helpful cheerleaders for the boss. Sometimes they are rebels and dissidents who sit on the sidelines working the crowd. Occasionally these informal influencers get to be leader if the position is handed to them on a plate and there is little risk attached. In highly communal organizations—as in some partnership firms, sports teams, and arts ensembles where everyone is of more or less equal status—the informal influencer may be persuaded to step up to the plate, as long as responsibility is held collectively or underwritten by some higher power.

Last, there are the Type Ds, the *reluctant leaders,* who want neither to compete for status nor to be looked to as leaders of people. In every society there are many who will opt for a safe and rewarding niche, if possible, free from competition. The attraction of this option is why the supply of leaders is limited. Whatever many people claim ("I coulda been a contender!"), they are actually content to be followers. So how individuals get to be reluctant leaders, for they are usually pretty awful when they do? It happens in one of two situations: either the supply of the other types is erratic or a particular leadership position confers little status or power.

An erratic supply of leaders is a feature of family firms where dynastic succession contains no guarantee that Junior will possess the motives that inspired Senior. Low-power reluctant leadership is characteristic of civil service and other bureaucracies where people are delivered to leadership positions via an escalator of tenure. Stick around long enough and you'll get to be the head of the department. Another bureaucratic model is to designate leaders by means of the carousel principle, as do university academic departments. Every few years the carousel stops and it's your turn to get onto the lead horse, whether you want to or not.

Ability

Drive alone is not enough. It has to be allied with skills. Leaders in whom the dominance drive is uppermost require above all else the skills to be able to influence others. If they lack these and have no capacity for charm, then they fall back on brute force. Al "Chainsaw" Dunlap made himself a legend of managerial toughness in various firms by delivering shareholder value through slash-and-burn cost cutting and downsizing. During his last job at Sunbeam, his grip remained secure so long as the company's share price held up, but without friends and trust there was nothing to save him when it collapsed. The culture of brutality he had championed turned its knives on him with little compunction.

Even his own adult children publicly rejoiced in their father's downfall. Robert Maxwell, the newspaper tycoon who died under mysterious circumstances in 1991, leaving behind him a trail of financial corruption, likewise inspired fear. Both men relied on their native intelligence and dominance. Maxwell could also turn on the charm when he wished. Both men had what might be called charisma.

In order to dominate a large corporation, a person has to be smart enough to have a coherent vision and to be able to deal with the politics of power. There is no need to be supersmart. In fact, it can be a liability to be too intellectually complex when what people want is a simple, clear vision. But you do need to be smart enough to be able to deal with the clever people you are likely to find around you.

Leaders driven more by the desire for status than for dominance require more specific talents. In an open competition for status one person succeeds by outdoing others in some domain. Bill Gates's obsession with winning and his willingness to bend the rules to get his way were legendary, but supported by his high intelligence and technical gifts. Many a corporate leader gets to the threshold of the top job by virtue of his professional skills, though at the last hurdle the winner usually gets to the top spot through political gifts. John Gutfreund, chairman and CEO of Salomon Inc., the investment bank, was a brilliant bond trader whose political skills got him to the top slot. As it turned out, this was not enough. His defensiveness and indecision in the face of a crisis over illegal bids made by the firm in 1991 cost him his job.

Roberto Goizueta at Coca-Cola displayed high intellectual skills in his chosen profession as an engineer, which served him well for the first part of his career which was in a technical environment. Slower to emerge but more influential was his strategic sense. It was this, allied with his drive, that enabled him to make his mark by radically reorganizing parts of the business and introducing bold market initiatives. As he rose in power, his skill in building alliances became critical, even though he lacked natural flair as a communicator. When he finally got to the top, his vision and intelligence came together to give him one of the longest and most successful leadership tenures in American business history. His example shows that there are other paths to dominance than by the more flashy charisma exhibited by his famous Cuban classmate.

Constitution

The management literature is curiously silent about the physical qualities of leadership; it doesn't fit the ethos of modern management to draw attention to primitive universals in how we respond to the look and sound of each other. We might wish it otherwise, but it is a distinct advantage to be tall, attractive, have a good voice, confident posture, and imposing body language. Our responsiveness to these cues is instinctive and often unconscious. It is easy to be impressed by a talented actor with a striking bearing. Ronald Reagan, aided and abetted by a network of admirers and associates, used fine delivery and an imposing presence to make up for what he lacked in intellectual ability. Leaders know these things matter and attempt to compensate for any lack of them by paying a lot of attention to their dress, appearance, and posture, with the aid of expert coaches and makeup artists. To enhance her profile of command, Margaret Thatcher took voice training to replace her vocal shrillness with an authoritative sonority.

What, I hear you ask, about the Napoleonic leader, the small man or woman with big charisma? It is part of popular legend that such people have to try harder. If they do, it is not through platform shoes and theater training but through their relentless drive. For perhaps the most important physical attribute of all— the quality that marries motivation to ability—is energy. Occasionally lazy leaders get by: the elder of the tribe sometimes is just required to show up and behave properly. But for the most part, the demands of leadership in business cannot be withstood without stamina and persistence. These, as much as physical appearance, are psycho-biological givens.

VARIETIES OF LEADERSHIP

D RIVE, ABILITY, and constitution come together to make leaders. Charismatic leaders seek both dominance and status achievement. The most successful are able to conjure visions through their words and infect others with their confidence. They seem to understand and offer solutions to their followers' needs and wants. They inspire trust by showing they share people's values and concerns. They are imposing in appearance and demeanor, displaying a reassuring strength, stamina, and durability. These qualities don't just apply to the good guys: many of the greatest villains in history were charismatic.

Less spectacular leaders vary in these qualities. Quite specific aspects of drive, ability, and constitution may distinguish the valued but noncharismatic leader, such as:

◇ The first-among-equals leader of a team-based business or operation

◇ The technician or artist leading by example in some highly specialized organization

◇ The inspired organizational publicist running a highly market-oriented business

◇ The wise elder of the board overseeing the large, long-standing corporation

◇ The nurturing protector governing a family-type organizational community

◇ The warrior chieftain leading a firm to battle in a tough marketplace

Individuals rise to these positions by a quasinatural selection process, an interaction between the qualities of the individual and the selective pressures of a particular business environment that favor one type over another. Some degree of selection is operating even when leaders are handed the role by birthright or seniority. They can still step (or fall) aside because they lack the motives, ability, or constitution for leadership. Many dynastic leaders have tried and failed to groom their offspring to succeed them by intensive resourcing and the best that can be bought in educational, military, and business finishing schools. The result is a hollow leader: someone who knows the script but cannot act the role.

WHY LEADERS ARE BORN, NOT MADE: BEHAVIOR GENETICS AND IDENTITY

IT USED to be held that what is most interesting and different about a person is the result of their upbringing. "Give me the child until he is seven, and I will give you the man" was once the unifying creed of the odd assembly of behaviorists and existentialists who held sway in psychology from the 1950s to the 1980s. This was in keeping with the optimistic postwar ethos fostered by many intellectual refugees from Nazism. It was allied to the belief that we could create a new world order after the horrors of World War II. Since that time, ideas of inborn character have been too close to doctrines of racial supremacy to get any kind of impartial airing. As we enter the age of biology and the decoding of the human genome, it is time to return to the idea that some people are simply born with potentialities for leadership. But this is a notion that will continue to meet resistance from social scientists obsessed with the social construction of reality. This thinking inspires a two-part fantasy. First we engineer utopian environments, and then we shape the people to fit them, Pygmalion style.

Awakening us from this reverie are new insights coming

thick and fast about the genetics of human character development. Genes set the parameters for the three attributes of leadership we have discussed. Drives are anchored in inborn temperamental and emotional sensitivities, located mainly in centers of the midbrain. Abilities have preset biases and limits wired into the neocortex. Constitution is programmed into a range of hormonal and bodily functions. In some of these capacities you can see the working of inheritance—parents' good and not-so-good genes coming out in their children. But many of the most important leadership capacities—those having to do with motivation and personality—are heritable, yet don't run in families. This sounds contradictory, but it isn't. Here's how it happens.

Every child's genetic profile is a random mix of genes bequeathed by her parents. Imagine Mom and Dad as cardplayers, each holding an array of cards. This hand is the genetic profile of character-related attributes each of them possesses, the hand dealt them to play in the game of life. Suppose each of the two parents has been holding throughout their lives a pretty average hand. The successes they have had to date have been a mixture of good fortune and good play with their limited resources. Conversely, their misfortunes have been a combination of bad luck, poor play, and the weakness of their cards against superior opposition. Now comes the moment of parenthood—the point of conception when their two sets of genes intermingle. This is the equivalent of each parent removing at random half the cards from their hand and throwing them on the table. Together these make a new hand, which is picked up by their offspring at the point of conception. The child will have to play this hand for her entire life. Her genetic complement from her parents' union will be neither added to nor subtracted from in her lifetime. But lo and behold, instead of getting a mediocre hand from her mediocre parents, the girl has picked up a truly wonderful combination of cards—a straight flush in poker or a fistful of trumps in bridge. This is how Herr and Frau Mozart—not untalented people, but not exceptional either—produced little Wolfgang Amadeus, possibly the greatest musical genius of recorded history.

Of course, the reverse can also happen: exceptionally gifted parents can produce quite ordinary children. Mr. and Mrs. Superbright, who met and fell in love when they were both postdoctoral students doing advanced work in theoretical physics, are dismayed that their little Johnny grows up to be a poor student and an underachieving adult.

The genes of leadership follow this pattern. We shouldn't expect them to run in families, but they do have a genetic component. This is because the genetics of character are not simple like those of eye color but work through complex combinations of the cards each of us holds. Even so, some single genes are being discovered that affect elements of personality, via the brain chemistry of neurotransmitters. Key elements of shyness—novelty avoidance and detachment—have been linked with specific genes in this way.

Another clear pointer to the genetic basis of personality is the recorded similarities and differences between adopted non-identical (two eggs) and identical (one egg) twins. Especially interesting are identical twins separated at birth and raised apart. On being reunited, such twins are liable to find they have been smoking the same brand of cigarette, supporting the same political party, following the same sports, and reading the same books. And once they are reunited, when they start to send each other birthday presents, the same gifts cross in the mail.

BECOMING A LEADER—FROM POTENTIAL TO ACTUALITY

THESE AND other studies tell us that up to 50 percent of one's personality is inherited. So where does the other 50 percent come from? This is proving surprisingly difficult to discover.

Personality has relatively little basis in general factors of the family environment and more to do with the specific life experiences of the individual. This is good news for parents. You are not responsible for your child's personality! Equally, it is bad news for the parents who thought they could turn their Bill into a lawyer and their Beth into a doctor: Bill insists on becoming an artist and Beth a schoolteacher.

What seems to happen is that as we get older we grow to become more and more "like ourselves." Our personality biases us to prefer certain kinds of people as friends and certain kinds of activities for play and fulfillment. These reinforce the direction in which our psychic rudders are steering us. The process starts in childhood. Kids condition their parents long before they are aware or care about what their parents want of them. Age brings more and more choices, many of them irreversible, about what to study, where to work, and whom to bond with.

Roberto Goizueta's story illustrates how this unfolding occurs—a mixture of genes plus luck. Early on he declared his independence of mind by turning down a safe route to fortune in the family firm and instead took his chances as a junior executive at Coca-Cola. His character showed itself again later, when he could have quit for the many seemingly better job offers that came his way, and in his ability to maintain critical alliances during the risky period of the run-up to his selection for the office of CEO. Once he was in the top job, he built the connections and structures that would support his strategy, as all great leaders do. Many businesses, in fact, exist only as extensions of their founder's personality and vision. People such as Richard Branson of Virgin would probably never have survived in a conventional business, such as his archrivals the conservative British Airways, but he functioned fine in businesses built around his own personal style and values.

Personality, via emotional reactions, also guides our development by telling us when we have made mistakes. Many youngsters make false starts in their careers (often because of pressure from parents and educators). It then takes a succession of job experiments for them to discover their true preferences and life path. Eventually, with the right opportunities, through a long line of choices people home in on a matrix of work, love, and leisure that suits them. This may take a lifetime. Many people, because of irrevocable early choices, get mired in jobs they hate, marriages without love, dreams they can't fulfill.

Everyone makes compromises. Some parts fit better than others, and many leaders sacrifice some aspects of their lives for their careers.

JOHN'S STORY

John is born to an average middle-class family in the Midwest. Early on he displays mildly obsessional tendencies. His absorptions are unlike those of his peers. He does not have a comfortable time at school, but he attracts a small coterie of friends. For them, John is the best game in town. Things happen around him. But John is more fickle than they are. He seems to switch off and on, spending a lot of time on private projects that he only occasionally shares with others. His parents encourage him in what he seems to be good at. He has various talents: He is good with mechanical things. He is greatly interested in money, and by various schemes he is able to do without his parents' handouts.

Without warning, at the age of sixteen he announces he has decided to quit school. His parents are horrified and try hard to dissuade him. But John is determined. He says he has a plan to make money, and all he needs is a loan from Dad to get him started. Dad eventually obliges. John and his pals set up a small operation that makes money startlingly easily. Before long he is running a business, commanding people and resources with assurance. Years pass. The business grows.

But his very success becomes a source of difficulty. Everything works fine as long as John is able to lead and direct all key operations and decisions. When things start getting too complex for him to control, he appoints one of his disciples to take charge of the daily operation of the business. This leaves John free to wheel and deal with customers and suppliers. It doesn't work. John isn't happy with this restriction on his role and keeps intervening in operational decisions. The situation is made worse by the fact that John's second in command is another techie with a poor touch with people. Between them, they confuse and upset the workforce. Morale collapses. People quit. Eventually the business fails. Some time later, John has a new business plan and starts another company with a new group of disciples. Eventually this too fails for the same reason—it grows beyond John's leadership capabilities. He is an entrepreneur—happy as long as he knows and can directly control the people and resources close to him.

John is like many people who start and run businesses: never content in any organization they are not in charge of. And when such people do run businesses, their vision is personal and their interest in conventional organization weak. The difference between success and failure for such people often amounts simply to whether they are fortunate in the resources and people they call upon. Many entrepreneurs instinctively recognize this and deliberately avoid growing their firms to a size that demands they become bosses of a conventional firm. Some strike it rich with a product and people that make their business a world-beater. Many highly successful firms had such founders. They remained terrible managers but inspiring leaders. Through the dedication of their more conventional colleagues, people who can cope with conventional organization, the business prospers. The founder remains a charismatic figurehead and inspiration, but you wouldn't trust him to run anything requiring organizational or social skills.

By the time you have reached your early twenties, like John, the dials on your personality profile have been pretty much set and so they will remain for the rest of your life. Readers in middle age may have had the experience of meeting an old friend after twenty or more years' separation. Once you've got over the momentary shock of the changes in your friend's appearance, a few minutes of conversation strip away the years. In a deep sense, this is plainly the same person you have always known. Even though their occupation and interests have changed dramatically, their essence is unaltered—you hear it in their way of speaking, comprehend it in their way of thinking, and sense it in their way of feeling.

It is this permanence of character that fits individuals for various roles in society. It is what makes some people natural leaders and makes leadership hard for others. You might wonder why evolution allows something of such fundamental importance as personality to vary so dramatically. The answer is that it is good to be different from other people. In the jargon of marketing, uniqueness of personality gives you your unique selling proposition, or USP, to prosper in the complex web of the human community. Being different gives you a brand to advertise when choosing a mate and talents to use in finding an occupation.

If almost everyone wanted to be a leader, the remaining few who liked being followers would make a good living, with leaders bidding in competition for their followership. Conversely, in a world of followers, the few who want to be leaders would clean up. On this logic our species contains different types, with all the attributes of personality spread across a population. (Chimps, incidentally, with whom we share 98.4 percent of our genes, seem to have much the same personality structure.) Personality variation allows us all to find a niche and exploit the value of our differences from one another.

THE NATURAL SELECTION OF LEADERS

LEADERS COME to power through both internal and external selection. The internal is self-selection, the force that induces some to seek and others to avoid leadership. Even when one is handed a leadership position on a plate, one's inner self has to acquiesce, especially around the three attributes. Are you ready to take on the risks and responsibility (drive)? Do you believe you have the skills to take on the challenge and succeed (ability)? Can you withstand the demands on your energies (constitution)? Many people have succumbed to the lure of the rewards or the encouragement of others and have become bad leaders, when they would have done better to heed their inner doubts. The self-doubt is likely to recur, and that can be fatal for leaders when the going gets toughest. But from time to time someone unlikely is catapulted into the top job and makes an unexpectedly great success of it when his or her attributes have a chance to be revealed, like Ricardo Semler's transformation of his family business in Brazil, recounted in Chapter 1.

Individuals who become leaders, whether by accident or design, get a helping hand from biochemistry. People who elect to take leadership positions are more likely to display the alpha male biochemical profile: elevated testosterone (the dominance hormone) and serotonin (the happiness hormone). More significantly, the levels of both, but especially serotonin, also rise *after* one acquires power. Leaders start to radiate health, confidence, and energy. They become more attractive. This is nature's way of giving status advantage a helping hand. It also carries the risk that the euphoria of dominance will blind incumbents to their limitations in other areas. It is the drug that makes leaders cling to their jobs long past their expiration date.

The external side of natural selection is the ecology of organizational life. To understand the leader's successes and failures, one needs to consider the many different situations in which one can lead. As we have noted, being the project leader of an egalitarian team is quite different from heading up a monolithic hierarchy. Who gets to be a leader is not just a matter of self-selection, except for people such as Richard Branson and Mary Kay Ash who create their own businesses around their identities.

Most leaders are *chosen,* and therefore we should pay as much attention to the process of choice as to the qualities of the leader.

In every community and organization there are three basic ways of selecting a leader. The leader *emerges* from the group by the organic process of competition or consensus. The leader is *selected* by the community because of some attribute or quality. The leader is *designated* because he or she meets some criterion or condition. Sometimes all three ways are linked together.

> Victor, by general admission the dominant force in his company's marketing department, is selected to replace the departing head of the group. Subsequently he is chosen by the board to take over as CEO. Soon thereafter, Victor announces sweeping changes in top management—demoting and sacking some key people and replacing them with his allies. He also announces a change in the appraisal and succession system. This turns out to favor more people from the non-finance-based side of the business.

It is just this sort of process that leads to the transfer of power from one kinship group to another in tribal societies. An individual's emergent qualities are widely recognized. This person is then selected for a position of responsibility. He then engineers change in the selection criteria to ensure that future leaders are designated from his own kinship group.

In business, reliance on any one of these three methods has both advantages and dangers.

Emergent Leaders

This type of leader arises in teams where people know one another and are able to determine which person they have confidence in. It is the organic consensus that enables a hunting party or musical group to know whom they want to lead them. The danger in this type of leadership selection is that highly dominant individuals may hold sway against those with other talents. Emergent processes predominate in highly political firms. Here informal networks rather than systematic review of talent decide who gets the sought-after leadership positions.

Where a leadership position offers great prestige and high compensation, the emergent leadership processes can stimulate too much competitive striving at the expense of other qualities. But where an emergent leader gains no extra power or special reward, just extra responsibility—as when one is picked to be the foreman of a jury or the captain of a sports team—it can lead to diffident leadership.

> Lars is from Sweden. He is the managing partner in a highly successful consulting firm. The firm is growing fast, and so are the burdens on Lars to pull the firm together. The firm's employees need a leader. But Lars is not a people person. He likes to be dominant, but he is not sensitive or nuturing. He is a tough guy. At the same time, he is open-minded and hardworking. I asked him why he had taken on the role of managing partner. His reply was that it was an accident of company history, not his or anyone else's active choice. In the course of our conversation it became clear that this role was not what he wanted, and no doubt it was ultimately bad for the business. Lars as a leader was at his best facing outward rather than inward—negotiating contracts with clients or selling the business, not being a shepherd to his flock. Lars reflected on these issues and determined to go back to his firm to renegotiate his role with the other partners.

Selected Leaders

In tribal societies, public contests may be arranged to choose who will become the leader. Electoral processes do this. Chieftains with different qualities are selected for times of war and peace. The same thing happens in modern politics: different types rise to the top of the heap to match the demands of the moment. In organizations the burden of selection decisions at mid to lower ranks typically falls to the human resources function. It is their impossible task to serve the business's future as well as current strategic leadership needs through the machinery of appraisal and review systems.

This works quite well in some firms for moving talent through the ranks to where it is needed, but it also has major failings: unexceptional people get promoted; women and minorities are undervalued; dominant males prosper.

Annette is the human resources director of a large construction company. She reports directly to the CEO. He is a monster, a forbidding, moody bully. He has a powerful intellect, a physically imposing presence, and a gargantuan appetite for work. He holds in contempt, and shows it, anyone he considers to be inferior to himself. He leads by fear. No one likes him or trusts him except Annette. She understands and forgives him. He is entirely dependent upon her as the only person he trusts and respects. She is thus his link to the rest of the organization. Annette herself is emotionally tough, sociable, open-minded, warm, nurturant, and conscientious. One day perhaps she will be a leader. In our conversation she says she would like to be, but not in this business. Her instincts about this are surely correct. She will be effective only as the top person in a firm where she can be part of a web of positive relationships.

Annette would make a wonderful leader, but what organization can accommodate her? Most, without realizing it, have shut out the Annettes of this world by allowing their cultures to get clogged with selection sclerosis.

Selection systems for leaders get it right, of course, when they have real talent in the pool and they allow it to emerge. The trouble is that in the long linked career systems of large business organizations, talent doesn't always make it to the top. Gifted people get too impatient for a prolonged series of incremental steps. They can't afford to trust the capricious filtration of the selection system throughout the middle of their career, when other options are closing; it's better to quit and take advantage of any better opportunity that comes along. Selection systems also shut out other types: creative, visionary people who are unwilling to engage in local competitions on parochial criteria. Good people lose heart when they see inferior people choosing other inferior people.

All these blockages set up a vicious cycle, reducing the supply of leadership talent. Top management, seeing few people they rate as good enough, then begin to recruit externally for key posts. This is offputting for the potential leaders already in the organization, who then leave in search of greener pastures. Over time, the firm's profile narrows to a monoculture of mediocre look-alikes, increasing the likelihood of having to look for leaders outside the company.

Designated Leaders

We noted earlier how reluctant leaders—people with no desire for the role—are delivered to leadership positions via dynastic, escalator, or carousel succession, that is, by birthright, tenure, or rotation. It is curious, therefore, how common and popular designation is as a method. Since time immemorial, leaders have been chosen by kinship, seniority, or other near-random designation methods. To understand this, we should first remember that though reluctant leaders often appear by this route, designation will not always produce reluctant leaders. In fact, of all the succession systems, this is the one that yields the most variation. It is as close to randomization as you can get, so if there is diversity in the pool, designated leadership has the chance to throw up more wild-card leaders than other methods. Remember the story of Ricardo Semler in Chapter 1.

Designated leaders are also more acceptable when the concept of leadership is less individualistic than we now take for granted, as in tribal societies, and indeed, as in the more recent history of Western societies. Communities feel a collective responsibility for their leaders and leaders take on the mantle of stewardship for the culture. Max DePree, head of the furniture manufacturer Herman Miller, a family firm regularly on *Fortune* magazine's list of most admired companies, talked about his vision of leadership in these terms—as an act of service and a position of trust bestowed upon him for the period of his tenure.

Strong cohesive communities regulate their members, including their leaders, through shared beliefs and compelling traditions. This gives some latitude to wild-card variation and allows for exceptional leaders to come to power once in a while. This process is enhanced by the natural diversity of the community. Our ancestors would not have been able to move into and out of their tribes at will, in the way we have become used to as we move from organization to organization. Many businesses have benefited from the unpredictable talent shown by the designated successor, such as Tom Watson, Jr.'s triumph at IBM after taking over from Tom Watson, Sr.

We are in a different area altogether, though, when we move into "as if" territory—when a boss identifies a favorite "as if" he or she were a favored son or daughter. In the absence of the randomizing force of behavior genetics within the family, choosing someone "as if" he or she were a family member has a reverse effect: it encourages cloning. When a boss chooses his or her successor, he or she is not randomizing at all. It might look like designation, but it is in fact selection—the successor has to fit in with the boss's vision of leadership. This is usually someone like himself. Leaders are often narcissistic, and when departing leaders are allowed to play a part in the choice of their successor, radical change rarely happens.

GROWING A LEADER

THE IDEA of growing a leader should be handled with care. Growing is not the same as creating, as any gardener will tell you. The key attributes—drive, ability, and constitution—are pretty fixed by the time you reach adulthood. So what can you grow and how?

Growing your own means four things: good soil, good seed, good cultivation, and good harvesting. In corporate terms, good soil means a nurturant corporate culture. Good seed is the human raw material for leadership. Good cultivation means retaining and feeding talented people. Good harvesting means locating and promoting them. In a truly healthy community, talent selects itself and rises to the top. Similarly, the best people tend to develop themselves. They don't need a paternalistic organization to spot and water them, though they do benefit from opportunities to reveal their talent (such as challenging projects) and good role models.

Some corporations see growing as creation, the godlike ability to take raw human material and mold it into whatever is most wanted, including leaders. Substantial corporate resources are devoted to leadership creation. Is this money entirely wasted? No, because through the skillful investment of effort and resources it is possible to change people's *behavior*. But this possibility is limited by the raw material available to work with. A tree can be trained to grow in lots of interesting ways, and some material can even be grafted onto it, but a willow will always be a willow and never become an oak.

So companies can spend their money wisely and business education can do some of its best work by equipping managers with the knowledge and skills that will help them realize their talents. Make no mistake about it, key managerial skills—for example, how to communicate, organize, and process information—can be acquired through training. Indeed, communication skills are the most powerful and important tools in a leader's repertoire. But the possession of these skills alone does not make an effective leader.

If you throw resources at some hopeful in order to create a leader, it will either pay off because the person has leadership potential, or you will get the next best thing, a professional manager who does things right but never inspires or an endowed favorite, who looks the part but lacks conviction. If enough incentives are offered, people will simulate leadership drives without possessing them. This is a major cause of leadership failures in business, especially family firms. Overinvestment just turns the reluctant leader into the empty leader.

THE EP LESSONS FOR LEADERSHIP

LEADERS HAVE two main functions in the life of a community. One is helping an organization define and achieve its purposes; this is the leader's role in formulating a challenging strategic vision. The other is embodying the spirit of the community and helping to hold it together, to represent symbolically the identity of the group. The balance between these demands changes over time. At some times the priority is the mobilization of the group to meet some challenge. At other times the primary need is for internal divisions to be healed. In the healthy community these go hand in hand.

Leadership that fits this model is flexible and organic. But how can we achieve this ideal when in business organizations we have moved so far away from the communal ideal of our forebears? The answer is, we will always do worse than we would like to, yet from time to time we will be surprised and delighted by our leaders. There are some things that will enhance the supply of good leaders in business.

◇ It's okay not to want to be a leader. There are many other useful things to be done in positions of seniority. Just make sure the people who do want to be leaders are not valued simply because they want the position.

◇ Remember what qualities are wanted at the top—vision, integrity, spirit—and make sure you do not let the tournament processes in the middle ranks deliver only competitive bullies. Don't let your selection system destroy or overlook diverse talent.

◇ Watch out for monocultures. Check that diversity is not being eliminated by subjective elements in the selection system and the politics of subgroup tribalism.

◇ Open up selection to foster diversity. Talent spotting is expensive, time-consuming, and unreliable. It's better to take chances with appointments and see how they work out. Since you can't anticipate all the future needs of your business, this will help to optimize your opportunities.

◇ Beware the risks of long-range talent scouting and don't throw resources at favorites. Instead, create opportunities that enable talent to reveal itself.

◇ Change your corporate culture to find and promote the leaders you want—environments in which quality is encouraged, mistakes are tolerated, different kinds of excellence are valued, and relationships are communitarian.

◇ Create a grow-your-own culture; avoid the not-invented-here syndrome. Deliver opportunities early. Don't make emerging talent wait.

◇ Don't reward dominance, but don't punish it either.
Don't expect some of the most talented potential leaders
to be team players, but do insist that they understand
and care about the needs of the community.

◇ Create partnerships at all levels, but especially at the
top of the organization. This means getting different
types of people working together to achieve balanced
functioning. Don't expect everything from one person.

• 5 •

Playing the Rationality Game:
The Wonderfool at Work

THE HARDWIRED DECEPTION
OF DUALISM

W E WANT our trains to run on time. We need our systems not to crash. We want to be able to plan from one day to the next with some confidence.

From time to time we do get let down, but mostly things work as they are supposed to. We build machines and on the whole they function well for us.

Now we have entered an era when the complexity of our creations is beyond our comprehension and control. The Internet has taken on a life of its own, even though it is driven by human purposes. In this it resembles another great and complex system that we don't fully understand and cannot control: the economy.

We keep financial and information scientists close to the centers of power, taking comfort in their confidence to make predictions even though their forecasts are sometimes wrong and they often can't solve our problems. We honor them as the new shamans of society. And like their forebears they are sustained by a magical philosophy—a fantastical set of stories about the world and human nature. It is called rationality.

Economists, especially, look at statistical aggregates and construct models that fit them. In their world we human beings think in straight lines, from one goal to the next. They see us as maximizing our own benefits, coolly calculating values, and making decisions so that benefits outweigh costs. In the real world our Stone Age minds, left to their own devices, typically reason from A to B via Q or some other roundabout route. Even when we are satisfied, we go on looking for new things to do. We work much harder to avoid losing something than we do to gain something of equivalent value. We routinely let costs outweigh visible benefits. Our feelings prove to be an unreliable guide to values or outcomes.

Rational systems and the people who believe in them are essential to taming the irrational beast that is man. No matter if we entertain odd theories of human nature. It is just part of a discipline we need to stick to to make things happen on time, be free from breakdown, and meet quality standards.

Welcome to Mega Corp., your nearest (mythical) metropolitan corporation, which we shall be visiting from time to time in this chapter. Top management at Mega Corp. don't like to think that the world might be full of chaos and happenstance. They dislike the idea that not only the people around them, but they themselves too, are not ruled by reason. It concerns them to think that success often has much more to do with brute force than brilliant calculus. It is unsettling to suppose that people have limited control over the course of events.

These ideas were less frightening to the chiefs of our ancestral tribes, who had a more accepting view of the world around them and mankind's place within it. Their shamans' job was to guide their community through the realms of mystery, spirit, and unpredictable change. Our ancestors believed in a single world of matter, mind, and spirit in which the healers and high priests administered magic, art, and ritual to give people a sense of meaning, purpose, connection, and hope.

In our present time we have learned so much about the

world that there is no way back to this view of our place in nature. The amazing march of science has opened our eyes, and our technological achievements have made us intoxicated with our own powers. Knowing so much about the world, we are also struck by our own braininess.

There has been an important accompaniment to this: the idea of *dualism*, which started with the ancient Greeks and has continued in one form or another ever since. It is a philosophical belief, an idea of great importance, but a falsehood. It is the idea that we are separate from nature. It comes in many versions: mind versus body, nature versus nurture, humanity versus the natural world. Evolutionary psychology takes us back to a unified view of ourselves as part of nature. Our psychology is part of our physical existence. Our culture is an extension of our psychology.

Since we started on the road away from our tribal wanderings on the savanna, we have built a world of organizations and control. Here reason rules, and the idea of dualism has proved to be a great aid to us in our grandiose ambitions. It supports our belief in our powers of control. It helps us feel confident and autonomous. So attractive and universal is dualistic thinking that one wonders whether it is a hardwired delusion—a mental model that helps us keep our primitive motives under control. So let us look a little more closely at what the brain is doing and how this affects the way we work.

THE MIRACULOUS ORGAN

THE MARVEL of the brain is that into a mere 1.4 liters are crammed 100 billion neurons, each with five thousand connections with other neurons, yielding an object of greater complexity than anything else in the known universe. When scientists look at it more closely, they see a very strange and rather soft piece of hardware. The wrinkled, spongy gray matter swims in a rich biochemical soup through which its connecting fronds grow and change. We tend to talk about the brain as if it were a computer, but its wiring is unlike that of any machine ever constructed. Unlike the hardwiring of chip technology, our thought process is fuzzy, nonlinear, and full of imprecision, approximation, ambiguity, and change. The brain doesn't calculate; it seethes and waves.

When you start to learn a new task, whole areas of the brain light up with electrical activity. Different parts sing to each other down its neural highways, until you've got the hang of the skill, and then it resumes its quietly efficient humming activity. We know where in the brain many key functions are located: the areas for language and vision, for example. But when we call on these by speaking or looking, not only do these parts fire up, but a symphony of activity swirls throughout the entire cortex.

The brain is an expensive piece of equipment to run, consuming a massive sixth of our energy. Evolution doesn't like waste. To invest that much design and running cost into a piece of equipment implies that it must be doing something pretty important for our reproduction and survival.

Well, one thing it's not doing is high-grade logical reasoning. Nor is it doing any kind of computation worth the name. It's also pretty poor at accurate recall. In fact, it isn't much use for straight-line thinking of any kind.

Much of what our brain does happens below the threshold of awareness. Look at people on a downtown street of a metropolis at rush hour any weekday morning. Miracles of coordination take place as people move smoothly and without conscious effort through crowds without touching one another or stumbling.

Another miracle: Stand in front of an audience, say of sixty people, and start talking. You will know from a single sweep of your vision who is looking at you and who is not. This is an amazing feat when you consider what a small signal someone else's gaze is. Moreover, if you are a practiced presenter you can do this without losing your concentration or breaking the rhythm of your thought and speech.

This is simple stuff compared with the great powers of intuition we possess. The human mind really comes into its own when it has to jump to conclusions, test relationships, devise stratagems, tell lies, recognize familiar things, and discern patterns.

In addition, it has one special trick that makes all this quite difficult for us to talk about sensibly: self-deception. We resent the idea that we are driven by our unconscious instincts. We prefer to believe we are rational, in control, open-minded, and well intentioned. We are programmed to believe that our choices are unconstrained and things happen for a reason. We find it hard to accept that people who do things we don't like can be intelligent and well intentioned, as well as the reverse—that people we respect are capable of inconsiderate or stupid acts.

What could be the purpose of such self-trickery? It is a key part of our programming that enables us to hold together a coherent performance—to have conviction in our relationships and courage in our actions. Throughout our history, our delusions and egotism have been vital assets. They help us make favorable impressions in our social world, and in the tasks we have to perform they help us keep up our spirits in the midst of uncertainty, hazard, and threat.

We live in a much less closed social world than did our ancestors. We understand our environment far better, and we have greatly reduced the likelihood of life-threatening daily hazards. But our psychology has not adapted and remains configured for our long-lost world. We have created a dazzling variety of artifacts for work and recreation that we struggle to keep up with, systems whose complexity are beyond our comprehension. We can marvel at the new towers and turrets erected by our imagination and ingenuity. And every day we stumble across a pile of rubble—something torn down by someone's arrogant folly or reckless error. We are truly wonderfools.

THE FALL OF BARINGS BANK

MUCH HAS been written about this startling case: a venerable two-hundred-year-old institution—Britain's oldest merchant bank—single-handedly destroyed by the actions of one young trader, Nick Leeson, who in 1995 accumulated losses of $1.4 billion. This is not an isolated case. In 1995 a trader for Daiwa Bank in Japan ran up losses of $1.1 billion. One year later, Sumitomo Corporation, the world's largest copper trader, disclosed that over ten years it had lost $2.6 billion, largely due to the actions of a single trader. Regulatory loopholes have been closed to prevent some of these events from recurring, but every day in large companies big chunks of company profits are needlessly thrown away by some individual's arrogance, thoughtlessness, and greed. In the treasury department of Electrolux, the white goods manufacturer, the first month of 2000 saw one of its German traders bite $28.5 million out of the company's capital through unhedged currency speculation.

Let us look more closely at the Leeson/Barings case, since the glare of publicity and published accounts of the events tell us much more about what actually happened than is usually disclosed by the embarrassed and defensive firms that suffer these calamities. To extract EP insights we need to look at three sets of factors: first, the man, his character, and his motivation; second, the events and how his psychology was swept along by them; and third, other parties and their reactions to what was going on.

The Man: Nick Leeson

Although made legendary by his actions and colorfully portrayed in movie and book, what is striking about Leeson is his ordinariness—a typical boy next door. The eldest son of a plasterer in a largely white-collar satellite town outside London, Leeson grew up with a psychological profile with four distinguishing features. One was intellectual skills: he did well but not outstandingly at school without too much effort. Second were social dominance and interpersonal flair: he was popular and a center of attention. Third was status motivation: he had a strong desire to pull up and away from his working-class origins. Fourth, he was low in conscientiousness and moral control. He was in the vernacular "a lad," finding it easy in his early work years to maintain a double existence as a respectable and dutiful clerk by day and a wild "tearaway" by night.

The psychological stage was perfectly set for what was to follow. Leeson was close enough to the visible wealth of the capital city to feel its lure and gifted enough to believe in his own potential to get a piece of it. At the same time, he could see the barriers before him: he had no college degree and lacked the advantages of birth or class that aided the careers of so many big City earners, people he quickly came to despise as his inferiors in natural talent. His early experience of restricted opportunity as a clerk at Morgan Stanley, another merchant bank, just pushed his nose harder against the glass of the candy store window. His move to Barings was an ideal opportunity. Its fledgling derivatives trading business and amateurish management presented him with a series of opportunities to circumvent the orthodox routes through the levels of the status hierarchy, first by legal and later by deviant means.

The Events: The Gambler's Nightmare

Leeson's ability as a quick, eager learner got him early successes in helping to sort out debts in Barings' new derivatives business, a complex and high-risk type of investment vehicle. Lacking the qualifications to trade in the City, his reward for saving the company a substantial sum came when he was handed the opportunity to trade out of Barings' new office in Singapore. Here, as an experienced settlements clerk, he was also allowed to run the back office, the posttransaction accounting process. He quickly became successful and popular in his new role. What started then as a mild form of deviance—parking unsettled losses in what is known as an "error account" for later settlement—soon graduated into more serious crime. Leeson starting concealing losses in a secret account he had set up, an offense that would get him fired by any bank. As his secret losses mounted, so did his risk taking. The ease of deception soon overshadowed his fear of consequences, and his confidence and risk taking escalated. Then, in a disastrous sequence of events, the markets moved against him and he accumulated huge losses. His response was to double his stakes to frighteningly high levels. Then the markets reverted, moving in his favor and wiping out his losses. Emboldened, Leeson returned to the fray with renewed vigor. Now he started falsifying his reports to the bank's head office in London in order to keep funds flowing to feed his strategy.

When an earthquake leveled much of the city of Kobe, Japan, in 1995, it also brought Nick Leeson down. It moved the market against his huge risk exposure. Now he lost all perspective and started to bet ever-larger sums, to the point of trying to prop up the collapsing Japanese stock market index on which he held huge one-way bets. It was at this point that the bloodhounds of Barings' head office began to pick up the trail of his illicit dealing. But by the time they reached his door, Leeson, with his unsuspecting wife, was on the run, while the bank tumbled into ruin behind him.

There are several themes in these events that we shall explore.

First is how humans react to levels of demand and complexity beyond their capacities. Feelings, not reason, take charge. By the end, Leeson was overwhelmed and running on empty. The wild excesses of his final deluded days in the market were the wild thrashings of a man whose emotions had overriden his reason.

Second is the apparent irrationality of human beings' quite different reactions to loss and gain. Like many gamblers before him, Leeson countered his own loss aversion by doubling his losing bets.

Third, we are all prey to control illusions, and the sequence of events in Leeson's misadventure encouraged false control beliefs to an extreme level. To the end, he believed he could pull out of the tailspin that eventually destroyed him.

Fourth, we are very naive statisticians. We have a very limited native capacity to understand numerical data. Leeson completely lost sight of the numbers. Although he was not dealing with the most complex kinds of financial instruments, they were still beyond his and most people's capacity for numerical reasoning. His thinking followed the human preference for simple categories, not any kind of proper risk calculus.

The Context: All Limelight and No Searchlight

Leeson sought and attracted a lot of attention. The world of trading is a very public arena in which big winners get big reputations. Leeson loved the limelight and cultivated it. His first audience was the local circle of the Singapore office, where he was appreciated as a sharp and helpful colleague, solving the problems of other people's errors and losses. As his profits grew, so did his reputation. In a short while he became king of the Singapore exchange, a legend for his boldness and profitability.

He also basked in the esteem of Barings' London management. Colleagues in London added their accolades and gloried in his profits, promoting him to head of the Singapore office in 1993. Within a year Leeson's fictitious trades were accounting for a substantial proportion of company profits, and through them came handsome bonuses for top management.

The spotlight on Leeson did not reveal his faults, it just magnified his apparent virtues. Management saw what they wanted to see, a vindication of their own strategic judgment. Their own success was too closely tied to Leeson's for them to entertain doubts.

The improbability of the size of Leeson's ever-growing profits and the oddity of the ever-larger sums he needed to call in from London to support his trading activity fell smack in the center of top management's blind spot. Even when an irregularity cropped up and auditors were sent over to check out his trading record, Leeson's plausible narrative gave them the reassurance they sought, deflecting them away from what was under their eyes.

This is a story of wish fulfillment in a climate of success. It is the dangerous faith we are ready to put into other people who act with confidence and speak with conviction. It is also a story of people—top management—who were blind to their own incompetence in a field they did not understand. Leeson's contempt for their stupidity spurred his deception, all the more so because of their seniority in social rank.

Like many other everyday calamities, the Leeson/Barings story is shot through with irrationality and self-deception. They come from the bundle of design features of the human mind that served us brilliantly in our ancestral environment but cause us no end of problems in today's hyperrational business world.

What would one have needed to be a successful hunter-gatherer? Quite apart from physical attributes such as vision and dexterity, you would benefit from an array of mental gifts and biases. Among these would be sensitivity to your own and others' emotional states and the ability to make intuitive leaps and make snap decisions in the face of uncertainty and change. A memory adept at filtering and sorting experiences into what is central to your interests and what is not would be important to keep you on track. In your spirit you would depend on courage and confidence to persist against the odds and win through in times of danger. You would also need prudence—the ability to recognize and avoid all situations in which you might lose a vital resource. Finally, you would have needed a range of social skills. These we shall discuss in the next chapter.

THE FEELING FACTOR

To TALK, as we have done, about emotions before reason could pull us into the trap of dualism: setting one part of ourselves against another. What this expression really means is that we are feeling animals and that everything that goes on in the mind is inflected by emotion, even when we are unaware of it. When we make what looks like a dispassionate choice based on limited information—Shall I buy A or B?—invisible feelings tug at our preferences, whether they are brands of pasta in the supermarket or equities on a stock market. Distant emotional impressions from images and memory traces that are beyond conscious recall steer our final choice.

Once our emotions are aroused, they take over our reasoning more comprehensively. Leeson, by his own account, suffered from a flood of contradictory feelings in the early days of his deception: the anticipated shame of letting his colleagues down by failing, the fear of being found out, the anxiety of carrying out a public deception, the tension induced by his high workload. The ascent of his mountain of fraud was motivated partly by the fact that each step onward gave him a bit more breathing space. And the more his strategy worked, the more insulated against the stress he became. But the strain of his secret life was beginning to warp his mind. Toward the end he became absorbed with feelings of persecution, delusional fantasies, and desperate magical hopes.

Most of us are lucky enough not to get into this kind of trap, but our emotional radar does send us vigorous warnings at every turn about dangers, real or imagined, which, of course, is exactly what we have it for. The brain's emotional centers have their own wiring direct from the senses. Before you have time to figure out what a sudden noise is, your brain has already flooded with chemicals readying you for possibly appropriate reaction—typically flight or fight.

When your conscious mind has figured out—a microsecond later—that the noise was just a door banging in the wind, your alarm systems switch off and allow your defenses to return to their customary state of armed readiness. As animals in a natural world, our survival depends upon our ability to detect threats instantly. Conversely, our positive emotions lead us toward those things that promise gratification, which is one reason it is perilous to try to rule out emotions in organizations.

The biological universals of our emotions can be summarized in a closely interconnected threefold sequence. First, there is the biochemical blizzard that occurs in the brain whenever a good or bad experience arouses you. Second, there are physical feelings: the warm bodily glow of something good, the icy cold of fear as blood leaves your extremities and rushes to your vital internal organs, or the flushing heat of rage as the blood is pushed outward or inward to where it is needed for fight or flight. This is emotion as action. Third and last are the signs about your feeling state you give off that everyone around you can understand. Darwin was the first to recognize that we humans, like all other higher species, have a universal set of facial emotional expressions that we are hardwired to emit and decode. They exist so that we can share the benefit of one another's radar reacting to both good and bad things. Emotional expression also enables us to regulate how we deal with one another. For this reason, with practice, like an actor, you can learn to control your emotional expression.

Emotions cause problems in business either because they are artificially inhibited or because they leak out when you don't want them to. Take the example of what most people rate as one of life's most terrifying experiences. No, not encounters with spiders and snakes, but public speaking. This is scary because it is potentially a situation of the highest risk in terms of human priorities: how we appear to others. If you do well, you get a good reputation and the immediate afterglow of people's good opinion (which some will share with you). Make a mess of a speech, and everyone is embarrassed, especially you. Many managers have lost sleep over the damage to the regard others hold for them through a poor performance in a public arena. For some, indeed, it does prove to be a career-wrecking experience.

The universal lexicon of emotions makes this an especially high stakes game. Every speaker is acutely aware of being continually scrutinized and hence the fact that any emotions signaling vulnerability and weakness are potentially visible. Actually, this fear is exaggerated since most of the audience is likely to be more absorbed with their own reactions than bothered about your emotions—that is, unless you draw them to their attention. This is where you may spiral into panic—the self-fulfilling fear that you will display your fear. This kind of nightmare would be relatively unknown in any community of people who live together and know one another. In our modern world the need for frequent display before strangers is a new human hazard.

In the closed community of tribal peoples, emotional expression is an act of sharing—whether rejoicing, mourning, or just getting comfortable around the campfire. Through sports, entertainment, and other media we express the same spirit, though it is often fleeting and many people in business work and live largely in emotional isolation. In the best working relationships, teams, and organizations, humor will be found. Evolutionists have pointed out that one of the vital functions of laughter is bonding—lovers and friends do it a lot. In the most cohesive teams, laughter is often to be heard.

But humor can also be used to wound. In fact, all emotions have a power dimension. We are all familiar with the boss who belittles people with sarcasm or indulges his temper with people who are too lowly-placed to be able to strike back. Many of the leaders we looked at in the last chapter—people like Goizueta, Gates, and Dunlap—have been legendary for their emotional outbursts. The more powerful you become, the less you need to worry about your emotional expression. Conversely, people of low status often get quite good at detecting subtle signals of the emotions of the powerful, as in the way some secretaries learn to gauge and manipulate their boss's mood.

The management of emotions at work is complicated by the fact that individuals are born with different emotional profiles. There are two aspects to this. One is the happy-camper/misery-guts dimension. Some people are relentlessly upbeat and cheerful, even under adversity—sunny children who became rosy adults. Then there are the serious ones, who as adults are perpetually pessimistic and dour, even at the best of times. The second aspect is emotional stability. Some people are perpetually jittery, sensitive, and variable. These are the people whose frame of mind it is impossible to anticipate before they walk through the office door in the morning. They are especially prone to feeling stress at work. On the other side are the emotional neutrals. Nothing seems to rouse them, and they seem largely oblivious of the feelings of others. At its extreme is the clinical condition of autism— a kind of emotional blindness that is often accompanied by high technical gifts, for example in math or music. It has been said, indeed, that in a mild form it afflicts many of the men who are obviously so much more at home with machines than people.

Emotional types color the organizational culture. People of the same emotional type tend to congregate together: unemotional engineers and computer scientists with things, emotional social workers and consultants with people. Put either group in charge of the other and watch their helpless incomprehension at how the other group behaves. These differences can be found within most industries. In finance, for example, contrary to popular belief, highly anxious and emotional people often gravitate to safe areas, such as accounting and audit, while more emotionally armor-plated people can survive in the hotter areas of trading and deal making. Each of these emotional types needs to be managed differently. Many organizations are unable to do this. Indeed, the rational ethos of many businesses deskills their managers emotionally so that they then have to be sent on training course to relearn Human Nature 101: how to understand the people they work alongside every day. In healthy communities there is visible emotional variation. People tell others what motivates them and learn about others' emotional profiles.

WINNING, LOSING, AND HOLDING ON

OUR EMOTIONAL and thought systems are designed to work closely together—to come to quick and accurate judgments about what's going on in the situations we confront. In the world of our ancestors, life was fragile. Someone who got an advantage tried to hang on to it. Someone who saw a risk of loss tried to avoid it. For someone who is clinging to survival in a world of continual threats to life, one slip below the subsistence line could spell death. On the other hand, a chance to get rich quick has great attractions.

In business, as in every other area of life, we continue to be dominated by these concerns. When faced with any decision, we always ponder three questions:

How satisfactory is the status quo?
What could I gain, now or in the future, by a given action?
What might I lose, now or in the future?

Although we like to think of ourselves as rational decision makers, much of the time our answers to these questions are unconscious and instinctive. Nick Leeson was on a roller coaster of semiautomated decisions, driven by events and impulses from moment to moment. He was never satisfied with the status quo. In fact, the better things got for him the more he wanted and the more he had to lose. It was the imminence of loss that governed his behavior, as it does for all of us in extremity.

The uncomfortable thought here is that any of us could be as foolish as Leeson, foolishly conservative as well as reckless, according to how we feel about these three questions. Many a company boss has wrecked his firm by sitting pretty in what he thought was a safety zone—until the market ambushed him.

RISK SEEKING FOR GAIN

In the boardroom of Mega Corp., a consumer goods business, the assembled executives are not making balanced judgments. The chief finance officer has been detained at another meeting. In her absence her colleagues are making rash decisions. It is the function of the financial disciplines of accountants and economists to protect us (and them) against instinctive judgments. Now, without the CFO's restraining influence, this group has embarked on a little spending spree. The board, under the persuasive influence of the marketing manager, is happily dipping into the firm's substantial war chest to find reserves of money for a new advertising campaign, but not enough to seriously dent the bottom line. Mega Corp. has had a good year. This new campaign will actually have a negligible chance of achieving its objective, but everyone is seduced by the vision of turning around the fortunes of an ailing brand.

IT IS by such means that companies quietly hemorrhage away their profits: disaster comes slowly and by degrees.

People often mistakenly talk about natural human conservatism and risk aversion. In that case, why do casinos do such good business? What are so many people doing engaging in dangerous sports, hobbies, and jobs? Why is the Internet buzzing with stock market day traders? Why are there so many business start-ups when only 5 percent will survive beyond a year or so?

It is quite easy to get people to take big risks. Here's the formula. Convince them that:

1. They stand to gain substantially.
2. Their downside is protected.
3. Their status quo is poor.
4. If something goes wrong, they will be able recover from it.

Usually all four of these conditions don't exist together. But just imagine you are out on the savanna. You and your family haven't eaten properly for days. You see a fresh deer carcass recently killed by a lion, momentarily unprotected. You can sneak up and snatch it, but there is a risk you will be chased away by the lion. You have a safe escape route, so your life will not be threatened. Your buddy is standing by and will come to your aid if something unforeseen goes wrong. Go for it! (Bad call. You and your buddy get eaten by lions.)

Young men are big risk takers because of these four conditions. One, they look for quick gains of wealth, notoriety, female attention, or whatever. Two, their downside is protected because they have little or nothing to lose in reputation and resources. Three, their status quo is poor—as yet they have no recognized status in society. Four, they have exaggerated faith in their powers of recovery—fueled by youthful levels of testosterone, they believe they can do anything. As his drama advanced, Nick Leeson took to using "Superman" as his password. He was hooked by points 1, 3, and 4 above—the prospect of a big status climb from a lowly position and a belief that his wits would see him out of any difficulty. Point 2 also applied in his early trading days; his downside was protected by the fact that he was gambling with other people's money, though later, when he was well and truly hooked, he ran real risks of disgrace, job loss, and possible criminal charges. More sophisticated traders spread their risk and avoid this problem. Many low-income people who dribble away their spare cash on lottery tickets fulfill all four points, especially 1, 2, and 3 (big gain, small downside cost, poor status quo).

The same principles are guiding career decisions in the depths of Mega Corp.'s management hierarchy.

> Daphne has been a management trainee with the company for the twenty months since she left college. The training program has given her exposure to various parts of the business. During her spell in purchasing she has become acquainted with someone in a new Internet-based company that supplies Mega Corp. It has offered her an insecure position in an area of potentially great opportunities for growth and wealth. She has decided to take it.

Big companies are always perplexed at the high turnover among their brightest and best recruits at around the two-year tenure mark. And where these young people go is often to smaller, riskier enterprises. Their first stop postgraduation is often a large company. One large, secure institution, college, is swapped for another, the corporation. This gives young people time to spy out the terrain, acquire some marketable skills, figure out where their best chances lie, and engage in a relatively low risk search from a position of security.

Note that it is always the most entrepreneurial who leave first and the more emotionally insecure who stay. This, of course, is one reason why the Mega Corps. of the business world are so often outmaneuvered by small, fleet-of-foot companies. They are magnets for the conservative. They survive largely by virtue of their massive bulk and market power, and to the extent that they are able to nurture and hang on to the talent of creative people.

The risk formula also supplies a recipe to persuade people to change their behavior at work. Set your targets to ensure that there is personal gain to be secured. Convince people that the status quo is lousy. Guarantee them safety and a way out if things go wrong. The formula also works to encourage innovation. Creative people—always inclined to be dissatisfied with the status quo and in pursuit of the big win of posterity—thrive in environments where they have psychological safety and freedom.

Of course, the danger with behaviors induced by these means is that you can be mistaken in your perceptions. The gains may turn out to be illusory. Your downside may be not as well protected as you thought. You may have undervalued the benefits of your current situation. Your powers of recovery may be less than imagined.

RISK AVOIDANCE TO MAINTAIN THE STATUS QUO

Some weeks later, Daphne's boss, Austin, also gets a job offer from a small supplier company. This happens to Austin from time to time, but he rarely looks twice at these opportunities. He has a mortgage, a car loan, and three kids all coming up to college age. He feels sure that Mega Corp. cannot crash. It is too big and powerful. Also, after twenty-five years with the company he knows its systems very well and has made himself indispensable to the function he is in (purchasing). He knows that most of the alternative outside job opportunities that come his way could offer a better deal, but they would be much less secure. New companies come and go all the time, and why should he put his career and his family's lifestyle at risk for financial gains that could be short-lived?

THIS IS all very well, but Austin may have made a serious miscalculation. Mega Corp. is a mature business living on its fat. Within a few years newer, more agile competitors will start to outflank it in its core areas. The process has already started, but so incrementally that only a few executives have even started to grasp the situation. It may take an accelerating decline brought about by a general recession in the market to bring this home to top management. By the time this happens, Austin, and others like him, will find themselves to be ballast thrown overboard as the firm attempts to keep afloat. Whether the firm survives or not, Austin's career is likely to come to an untimely end.

In the boardroom, similar reasoning has been at work. A couple of years ago, Mega Corp. was approached by another large retail business making tentative moves toward a merger. This would have made a lot of sense in terms of complementary assets and probably saved the business. But Mega Corp. rejected the deal. The board looked at recent years' results and concluded there was no real threat to the company's market position. They were also motivated by feelings of uncertainty about what their own personal futures would be in such a radically restructured organization.

What we are seeing is a variation on our earlier points. The recipe here is that people will avoid risk or resist change when they believe that

◇ Gains are uncertain.
◇ The status quo is comfortable.
◇ Change has a potential downside.
◇ They will lose control if they move into new territory.

Is it any wonder that corporations have difficulty getting employees to accept change? Most people are tucked up as deep possible inside their comfort zone. Even if this is not a wonderful situation, it may be a lot better than the alternatives.

The alternatives, too, may be misread. Under John Akers, IBM went through numerous structural adjustments in a vain attempt to keep pace with the outflanking maneuvers of new, agile competitors in the suddenly exploding PC market. The company was terrified of abandoning its comfort zone, the strategies that had made it one of the most powerful and envied corporations on the planet. What it failed to see was the downside staring it in the face. It took a change of CEO, the abandonment of many of its sacred cows, and a radical departure from its former strategy to move it back to a leading industry position. The IBM story still has a way to go.

RISK TAKING TO AVOID LOSS

Meanwhile, back at Mega Corp....

A couple of years have passed. There has been a downturn in the economy. The warning signs to Mega Corp. are now inescapable. The company is facing competitive threats on all sides. Its cost base has become cumbersome and inflexible. The market risks are increasing. Management talent has been seeping away. The board enacts a series of crisis measures—slash-and-burn cost cutting. The training budget goes first, of course. Then consultants are brought in to get serious about downsizing. At the same time, the board decides to advance the launch of a new product it has been considering for some years but always backed away from because of the risks. "It's now or never," the CEO says to his desperate band. Like him, they know they will be cashing in their pensions if something doesn't transform the company's fortunes.

On the fourteenth floor, across the corridor from Austin, sits his colleague Richard, biting his nails. Richard has practically given up doing any company work, for he spends most of his time surfing the Net for one of the dwindling number of job opportunities in his area. Today he thinks he's found something. It's riskier than a job offer he turned down eighteen months ago, but now he feels he can't afford to wait. Meanwhile, Austin is doing nothing. He is ten years older and less qualified than Richard. He joined the company as a clerk, working his way up the ranks. He appears to be in a calmer frame of mind and goes about his work as if nothing is happening. He is keeping his head down, trying to be inconspicuous and reliable. He is clinging on in the hope that he will survive.

WHEN AN animal is confronted by a threat to its life, one of two things happens. One is that the creature will freeze. This can preserve life under some circumstances, such as when the threat is unavoidable and undefeatable, like a hurricane. Immobility can maintain an animal's physical systems until the storm has abated. But where there is the remotest chance of escape via action, such as when cornered by a predator, it is best to throw caution to the winds and try to scratch and bite one's way out. In a threatened firm, for weak workers the threat is a hurricane to be endured; for resourceful ones it is a wild animal to be fought or fled.

So we can add to our rules:

◇   People will take risks to recover from past losses and avoid future losses.

But . . .

◇   Where losses are unavoidable, immobility is the best option.

Hope leads us to fight rather than freeze—as long as there is some possibility of escaping the hazard. Our aversion to loss is one of the strongest of our survival instincts in a dangerous world, and it operates more powerfully than our desire for gain. As tennis star Jimmy Connors, a notoriously fierce competitor, once famously said, "I hate to lose more than I love to win." Casino managers have understood this for a long time. The people prepared to take the biggest risks are those who are chasing losses. The great Russian novelist Dostoevsky, an inveterate gambler, knew all about it. He describes the phenomenon of loss chasing in his classic short story "The Gambler," written at high speed to pay off his own gambling debts! Desperate loss chasing is what Nick Leeson did in the final hours of his wild trading to pull out of his spinning nosedive into debt. The same principle has been observed in many species: the closer they get to the survival boundary, the more chances they will take to secure vital resources.

Managers in financial markets understand this powerful tendency in human nature so well that they try to drum into new traders the edict "Cut your losses and let your profits run." Gamblers' downfall is to do the opposite and follow their instincts: they chase losses and take profits too early. The same tendency dominates business strategy. The fatter and more successful a business, the less it takes market risks, while the new kids on the block with nothing to lose take the entrepreneurial chances. When a successful company gets into trouble, it starts throwing everything overboard—the good along with the bad—in a desperate scramble to recover.

Because of the overwhelming focus on controlling loss, traders and their managers often give insufficient attention to missed opportunities on the upside. These effects are strongest at year-end, when bonuses are calculated. Traders who have had an indifferent year start to take bigger risks in the hope of getting a late boost. Traders sitting on a healthy surplus for the year become cautious to protect their standing. They are in a game where reputation is the critical currency, and they are highly motivated to preserve it.

In practice, traders and experienced gamblers learn to set strict limits to force themselves to cut their losses. Sometimes when a trader is getting sucked into a vortex of mounting losses, a manager has to step in and force him to cut his position. This is a tough call to bear. Managers tell traders who have been burned this way to take a few days off. They hope that the experience will be easily forgotten. But it isn't. Unless it is soon counterbalanced by a success—like riding one's bike again after falling off—failure lingers corrosively in the memory.

FEEDBACK, FAILURE, AND THE
FIGHT FOR CONTROL

 HERE ARE some of the scariest thoughts that people in organizations have:

◇ People don't think well of me.
◇ I have failed in some way, I am not sure how.
◇ Maybe I am no good, really.
◇ What happens to me is beyond my control.

All of this hinges on feedback, the most mishandled explosive in business. In our ancestral communities you didn't need 360° appraisal systems to know what people thought of you. Failure was up front, in your face, and real, not a theoretical abstraction. Self-esteem was not a lonely room in an impersonal city but part of a common space with others. Control was imperfect, but there were people around who would help you out if things got difficult.

In business bureaucracy feedback has become routinely mishandled.

Margaret prides herself on the fairness and accuracy of her judgments about her staff. She does an annual review with Bill in which she provides what she thinks is a balanced and honest assessment of his strengths and weaknesses. He says little during the interview, but she is shocked and hurt by the sullen negativity of his manner the next morning when she stops by his desk for a chat.

What may have gone wrong? The problem is the programming of our minds to have asymmetrical reactions to good and bad news. Margaret thought she was giving balanced feedback. There is no such thing. Just one drop of criticism can sour the milk of human goodwill. Her feedback was general and non-specific, and Bill was left with a sense of having failed without knowing why. This made the situation one in which he felt he had little or no control.

Failure is the F-word of business, probably its most taboo term. See how people use or avoid using the word, knowing it to be dynamite. We never apply it to our friends or loved ones unless we intend to wound deeply. We use it about our enemies, usually to third parties—never face-to-face unless we are in a big fight with them. We use it about ourselves to express our deepest shame, and if we keep on doing so we are probably in a state of clinical depression.

The idea of failure strikes at our most vital interests. Among social creatures, failing publicly, or being seen as a failure, is one of the most threatening kinds of loss. The need to save face is compelling. This and other extreme reactions to failure are born of a powerful instinct: the need to keep control over your vital interests. So extreme can be this drive that it may lead to self-defeating behavior—like that of the senior Japanese executive who one day in March 1999 walked into the CEO's office and committed hara-kiri in front of him because of a business failure for which the excutive had been responsible.

In research at London Business School, David Cannon looked in-depth at people's reactions to business failure and came up with some surprising and disturbing conclusions.

The first surprise was the lasting potency of emotions about recollected failure, even long after the event. When people recounted the events, they relived the feelings. Men and women alike cried openly, even over long-past and seemingly quite minor events.

The second surprise was that even when they clearly had not been at fault for what had happened and others had told them so, they still felt responsible. One example was a pilot who had damaged a plane in a difficult emergency landing. He had been commended for doing the best he could under the circumstances, but he continued to blame himself long after the event. The reader may recall the opening story of my mugging and my curiously "responsible" feeling after the ambush. This sprang from the same basic urge to reassert control.

The third surprise is that many people seem not to learn much from the experience of failure. Instead they engage in magical thinking. They say to themselves: Next time, I'll be careful so it can't happen again. Next time, I'll see it coming. Next time, I'll be perfect. This is the mind desperately trying to regain a sense of mastery. Blaming yourself for something you actually couldn't have avoided is paradoxically reassuring. It is less scary than the idea that you have no means of controlling danger, uncertainty, and difficulty.

It has become a mantra of management writing about the "learning organization" that we can and should all learn from our mistakes. This is well and good; the idea is a cornerstone of scientific method. But in R&D no one gets rewarded for instructive failures as much as they do for simple successes. Failure avoidance and the control instinct are a lethal combination in our complex world because they oversimplify. Instead of learning from mistakes, people go into defensive routines. It is more important to save face than to determine the causes of the failure.

Our ancestors believed in magical forces as a way of coming to terms with the limits to their control. In fact, such beliefs persist widely today. In the rational world of business there is no room for such nonsense, so we believe instead in the magic of leaders or consultants. Then if things go wrong, we can blame them.

CONFIDENCE BEFORE REALISM

YOU MIGHT be wondering how not learning from failure could be permitted in the human design. The answer lies in the dilemmas that humans have faced since their beginning. A recurrent theme of this book is that we evolved in a world that was simpler but more perilous than the one we now live and work in.

In such a world, in which losses are highly tangible and the status quo cannot be easily maintained, taking chances is necessary. The best way of taking risks is by not believing in the full extent of their riskiness. A good strategy for doing this is to believe you can take control of whatever happens.

We like not only to feel confident but also to work with and for people who exude confidence. Confidence is meat and drink to the leaders we looked at in the last chapter. It doesn't just make people bold; it also makes them persist in the face of great obstacles. The higher someone's self-esteem, the more inclined they are to try and try again. This is fine if persistence pays off, as it often does. The management literature is full of heroic tales about strategic successes, such as the persistence of Sony's boss, Akio Morita, in developing the Walkman or Walt Disney's seeking to develop an entirely new medium for film. Many people go to their doom still fighting furiously, and we honor and praise them for it. The high regard for hopeless heroism in our culture reflects the values of our mental architecture.

We hear less about the downside of confidence, but it is real enough and all around us. Because it helps people make light of difficulties, it can also be a path to disaster. There is an old joke that goes: the difference between an optimist and a pessimist is that the pessimist is in possession of *all* of the facts. It is no accident that people who are classed as clinically depressed have been found to possess a more realistic appraisal of life's risks and hazards than so-called normal people. It is estimated that more than half of all business decisions fail, and the confidence illusion is a major contributor. Psychological experiments show how prone people are to magical control beliefs. In financial markets, where you might think rationality would reign supreme, such beliefs are rife. There are traders who wear their lucky socks to ward off failure and investors who believe they can outguess the aggregate effects of millions of random market movements. And at the furthest extreme you have Nick Leeson and other rogue traders, held tight in the grip of control illusions.

A second downside of confidence is the way it overshadows other qualities. We respond so warmly to the contagion of others' confidence that we overlook other flaws. Media tycoon Robert Maxwell charmed and bullied his way through his business dealings, followed by a flock of dependent underlings all the way to disaster—death for him and ruin for his business. At a more mundane level, in many a selection decision one finds the inferior but confident candidate landing the job.

The third downside of confidence illusion is the negative effects of persistence. We tend to remember and celebrate the successes of those who tried and tried again; we are more inclined to forget the innumerable useless projects that have been vast sinkholes of public and private funds. For every inspiring success story in business history, there are a hundred cases of companies that vanished because their bosses stuck to a once-successful, now-obsolete strategy.

I conduct a classroom exercise that illustrates the effect. It is a case in which a company sets up an R&D lab but so mismanages the operation that it ends up with discredited management confronting a disaffected workforce with threats of disciplinary sanctions. I invite my students to take over the management of the situation. They do so with gusto, offering lots of ingenious suggestions to turn the situation around. Most of these proposals vastly underestimate the difficulties, and involve huge costs.

When I suggest it might be better for the firm to cut its losses and outsource the whole R&D operation, I regularly encounter a fierce negative reaction from students.

Some of them seem personally insulted by the suggestion that this might be a situation beyond their capacity to turn around. This reaction seems to be felt most strongly by young MBA students, a smart, successful, confident, and privileged group. As an elite group who have won through to a top international business school and are looking forward to a glittering business career, the last thing they want to be told is that there may be situations beyond their control or willpower. This message is more readily accepted by older, more battle-scarred executives.

The same persistence effect almost eliminated the mighty Ford Motor Company when Henry Ford maintained his one-model strategy while General Motors was becoming ascendant with a diverse product range. The same sentiment nearly destroyed IBM when it adhered to its once-profitable concentration on mainframes while Apple and others surged into the market with PCs. In these and many other cases, the underlying cause is the same. Success fuels confidence. Confidence makes leaders shortsighted. Warning signs, usually about how the business environment is changing, are ignored.

Research shows that people with big egos carry on with tasks long after they should have quit. It also finds that when you feed them accurate warning information they are much more likely than other people to ignore it and carry on. The inflated ego of the successful boss deepens his commitment to the "one last push" and "if it ain't broke don't even look at it" mentality.

This has always been true: the more puffed up with self-belief we are, the more likely we are to put ourselves and others at risk. The qualities we look for in executives and leaders are also those that imperil us most.

In the words of W. C. Fields, "If at first you don't succeed, try, try again. Then quit. There's no use being a damn fool about it."

TRICKS OF THE MIND

So FAR we have been talking mainly about how our motivation affects our decision making, which we know is located in the older, more primitive parts of the brain that we share with most other mammals. But what about the higher centers of mental functioning? Don't these allow us to be rational, efficient decision makers? Yes, we are very smart in many ways. We invented mathematics and use it to great effect. We created the computer and other machines to help us solve complicated problems. But the marvel of these things is just that: *we created them.* We did so precisely because what they do doesn't come naturally to us. Without their help our powers are quite primitive.

Can you multiply 78 by 67 in your head? If you can, well done. This is an easy computation, though most people can't do it without the aid of pencil and paper. Add a few digits, and not even a math genius can do it in their head. The brain does contain a hardwired number capacity (in the parietal lobe), but it is laughably puny. It enables us to assimilate instantly only a small number of unrelated objects around the number 7—just enough to determine at a glance if one of your family group is missing.

Even feebler is most people's grasp of probabilities and how to estimate them. Specialists are not much better. Studies have shown that surgeons advising on the likelihood of treatment outcomes give hopelessly wrong answers when asked to calculate the risks of a surgical procedure with the aid of some simple actuarial statistics.

Our incapacity extends to logical reasoning. It is very easy to tangle up people's thinking with only slightly complicated sets of arguments. Politicians, lawyers, and advertisers do it all the time, and not always consciously or with intent to mislead. We regularly fool ourselves with our faulty logic.

Why has evolution allowed this to happen? What can be good about being irrational? Surely it cannot be functional for survival to fail to use the great power our brains must contain? What else are we doing with all this gray matter? Why would hunter-gatherers have needed such a big brain?

On the other hand, if we are so bad at computing, why did it take more than forty years of hardware and software development for IBM to build a machine of sufficiently prodigious capacity, "Big Blue," and to program it cleverly enough to beat the world's best human player, Gary Kasparov?

The reason is that our big brain is designed to do things that computers find very difficult. Even people with poor memories can outplay quite powerful computer chess programs. The gift lies not in memory or thinking power but in strategic capability. The components of this are pattern recognition, planning, deception, and flexible action—not so much different from the skills of a hunter. We do have superb memories, but not in the way a computer does.

At random, pick a working day around a month ago and try to remember what you did that day. Unless something really significant happened, you will probably have difficulty remembering anything about it. Now give yourself a clue: look at your appointment diary. Now the memories will start to come back, but not as a blow-by-blow account of how the day passed. What you will remember is an uneven series of more or less memorable episodes and not much else in between. Anything that had a sufficiently positive or negative emotional tone will stand out: something you did that made you feel good; something that amused you; someone who annoyed you. You will also remember any important outcomes: decisions, things that were said that might have turned out to be important.

This is how memory works. It is active, not passive. We don't catalogue a library of records. We spot resemblances, see familiar connections, and reconstruct past events. We are tuned to recollect anything our emotions signal as having positive or negative value. Memory instructs us how to keep a positive emotional balance sheet on future occasions. We recall events as stories in ways that dramatize our own role, usually favorably. We remember little things that may have future personal relevance. And in recollecting it is rare that we merely indulge in the reverie of replay. It is hard to stop oneself from thinking about the future, considering the significance something that happened may have in your life ahead. We spot connections between memories and future possibilities. The act of remembering turns into emotional pathfinding, purposeful planning, and creative exploration—all vital survival functions.

But when we need to remember facts and things we can do so excellently with two aids. One is narrative. Put something into a story, and we remember it. The other is classification. Our brains are constructed to be able to think in terms of nested sets of categories. We are truly excellent at dividing things into categories and classes—seeing similarities and differences. We love lists; I have used several in this book, as they help us notice and remember things. Our memory for facts, stories, and objects is prodigious. Anthropologists have been amazed at the superb and accurate knowledge nonliterate tribes have of the flora and fauna in their environment. We are brilliant at spotting changes in the appearance of things and finding hidden objects—the kinds of abilities people need if they depend on foraging for much of their food supply.

Another curious feature of memory is that it is often lodged in the senses: we "see" scenes and "hear" dialogues. This ability can be trained. Doctors read X-ray charts with educated eyes and engineers listen to the hum of engines with clever ears. They can detect things they would be unable to explain or verbalize easily.

We are superbly equipped for a particular lifestyle: following herds, finding edible fruits, roots, and berries, anticipating and adapting changing conditions, and living successfully with our own kind. Our ways of thinking include various natural illogicalities—what psychologists call cognitive biases—preset ways of thinking. These served us well in a preliterate age of tribal living, but now they carry penalties in the modern world. Here are some of the most important.

Attention Biases

As technology marches on, one of the most difficult challenges our Stone Age minds face is how to deal with the ever-rising flood of information. You take a week off work and come back to find your computer clogged with e-mail and your in tray stacked high with reports. Go abroad, and it's worse: you have all the local business news to catch up on.

Luckily, our minds have natural filters that sieve all this for us. Unluckily, these filters don't always pick out what really matters to us. Our attention has several kinds of bias. They are well known to all skilled negotiators and salespeople, who use them to steer our choices toward their own objectives.

One is a kind of village mentality. It makes us pay attention to whatever is familiar and might connect with our interests. This is the indirect cause of many a business failure. Executives scanning their markets notice what is happening to the companies in their field and the businesses run by their friends and enemies but overlook new and unfamiliar threats emerging under their noses.

Another attention bias is the sequence effect. Our minds are adept at putting down markers for the sake of comparison, setting benchmarks to check whether things are getting better or worse. We also remember our most recent experiences most clearly, since they are likely to be most relevant to our interests. The combined result is that in any sequential presentation of information the first and last items tend to stand out in our memory. Go to a trade fair and look at fifty stalls; see which ones you recall later. Most will be a blur apart from the first, the last, and a few others that had something different about them that made them noteworthy. Job seekers, beware! Try to get onto the interview list as first or last in line; if you can't, make sure there is something special about the way you look or what you say that will really make you stand out from the crowd.

Sampling and Statistical Errors

The idea of randomness is alien to creatures moving purposefully through a world of living creatures, changing seasons, and willful fellow humans. It is scary to see how easily we can be fooled by data. If a bunch of numbers are thrown at us, we discern patterns even when there are none there. Presented with disconnected information in a sequence, we see a trend. Ask us to guess the true probability of either a rare or a common event, and our estimates are wildly inaccurate. Markets are being moved by legions of individuals practicing faulty calculus. As natural statisticians we are truly awful—which is why, of course, statistics is a mathematical discipline requiring extensive study. It is also why we use computations to help us make key decisions. Deep down, we know that we trust our "natural" judgment at our peril.

Just such misplaced trust is what killed the seven astronauts on the calamitous launch of the *Challenger* space shuttle in 1986. Nineteen previous successful launches had made NASA's flight directors overconfident, even though a twenty-to-one shot is lousy odds for a life-and-death gamble. Sampling theory would have told them how great the real risk was, especially when the decision makers held in their hands statistical data about possible O-ring failure—the very event that doomed the shuttle.

People often discount a risk until someone they know falls afoul of it, and thereafter they exaggerate its likelihood. This applies to entrepreneurs taking market risks, just as it does to smokers risking cancer or motorists driving at high speed. Our natural calculus is biologically geared to our hopes and fears. This shows in our tendency to overestimate the likelihood of low-probability risks (being in a plane crash, losing all our money in a stock market crash) and underestimate the likelihood of highly probable risks (being in a car accident, losing money on the stock market). Our fear—loss aversion—makes us very interested in unusual and rare occurrences. But our hope—our need to pursue goals with confidence, not in a constant state of anxious paralysis—makes us blasé about the hazards we have to live with daily.

Logical Reasoning Errors

Our nonlinear way of thinking, which is great for quick, intuitive action, gets snarled up when we are required to make rational choices or analyze causes. The illogic of people's preferences has many politicians and marketing executives tearing out their hair in frustration. How is it possible for the buyers of toothpaste to prefer Dazzle to Sparkle and Sparkle to Twinkle, but then to like Twinkle better than Dazzle? Why, when Twinkle is put onto the market, does it affect the sales of Dazzle and Sparkle in opposite directions, and when it is later withdrawn the sales of neither of the other brands rises? There is a lot of complicated theorizing about these and similar effects. The explanation is that we do not make comparisons in a logical or orderly way. We see an array of elements and then form an impression. Change one element and our impression of all elements changes. No matter that this defies the laws of logic; our minds are designed to make snap judgments and get on with the show, not waste time on precise calculations. Anyone who has ever drawn up a list of pros and cons when trying to make a tough decision knows this. The decision is never made based on the length of the two lists, but on the gut feeling you have about specific elements in either of them.

Our logic also lets us down when we are trying to understand why things happen. Superstitions are among our most common irrationalities. There are some that are obvious, such as wearing a "lucky" jacket to bring us good fortune in a job interview. We are less aware of many others. An executive who schedules a product launch at the same time of year as a previous successful launch is indulging no less in superstition. So is a marketing manager who claims that a rise in sales is the direct result of a new advertising campaign even when the whole market is booming. Superstition—irrational causal beliefs—affects almost every important decision we make where conditions are complex and outcomes difficult to predict. It is a variant on our theme of confidence before realism. We prefer to believe in *something* rather than to be unsure. Take executive appointments, for example. Because it is very difficult to identify what will actually make a given candidate successful or unsuccessful in a job, interviewers base their judgments on visible factors, many of which turn out to be irrelevant: the college someone attended, where he is from, the way she speaks, his appearance. A variant is the tendency to overgeneralize from a small sample. It takes only one or two women to fail in a male-dominated organization for their sex to become the theorized cause of their failure.

Social Biases

Almost all our biases are attuned to our social existence in one way or another. Whatever happened on the savanna, the most important day-to-day events involved other people and what they were doing. So we have an inborn tendency to tune in to people before things, people we know before people we don't know, and people like ourselves before people who are different. If we are in doubt about what to do, we watch what other people are doing. Go to any meeting where a vote is taken. People with clear, strong opinions will wave their hands in the air as if claiming leadership. Others, less certain, will watch and go along with the people they like and respect or just follow the majority, rather than abstain.

This is a well-known psychological phenomenon known as herding. It has two drivers. One is finding the safest place in the herd, which is in the center, as it is the weaker members that straggle along the fringes and get picked off by predators. In many social situations we do not like to stand out, preferring to feel secure in a throng of others. This is part of the reassuring pleasure of being a spectator in a large crowd. The other aspect of herding is imitation—the tendency to follow the majority rather than figure out for yourself what to do.

Herding works both consciously and unconsciously in all kinds of decision making. It is what leads people to pass empty restaurants by. It causes shoppers to buy known name brands and firms to choose well-known rather than equally good but obscure suppliers, advisers, consultants, and other professional services. The logic is circular. Small suppliers are forced out of business because of prospective clients' reasoning that they can't be any good if other big firms don't use them.

If people are challenged about their herding behavior, they are likely to protest that they are making a rational choice on the basis of objective attributes. But how different are we from our creature cousins when making difficult choices? In one species of grouse females appear to choose which male to mate with apparently on the merits of his display, so that one lucky male gets to account for more half the impregnations in a given group. Closer observation reveals that the females have actually been watching and following the choice of other females. If you are trying to make a close call—Is Dazzle really better than Sparkle?—why waste a lot of mental effort for a marginal payoff when you can simply buy the brand that most other people seem to prefer? Something similar seems to happen in markets. Exchange rate movements are almost completely unpredictable, but investors pile into and out of currencies in droves, creating largely mean-ingless "trends." The performance rank order of mutual fund managers changes dramatically from year to year as investors chasing profits move in droves from last year's favorites to this year's.

Yet imitation makes evolutionary sense. It is better to base one's own choices on the aggregate wisdom of others than to trust one's own judgment—especially when the stakes are high and the penalties for being wrong are heavy.

THE IRRATIONAL GIFT:
FOSTERING THE CREATIVE SPIRIT

FAR FROM being the enemy within, our irrationality is merely a gift out of place. It is only because of our emotionality, persis-tence, overconfidence, personal biases, number blindness, illogi-cality, and all the rest that we are so wonderfully inventive.

Rationality gets machines to run. Intuition invents them.

We are great at guesswork. Our ability to make intuitive leaps is remarkable. It is so subtle and complex that it is beyond our conscious comprehension. The seasoned country dweller does not know exactly upon what evidence she draws when she looks at the sky, sniffs the wind, and foretells rain. We read other peo-ple like the weather—because our lives may depend on how well we can predict their actions. We "sense" social situations with the same kind of intelligence. How often have you walked into a meeting and "known" that people have been talking about you, or intuited without any obvious signs that something is wrong with a close acquaintance? We have needed these gifts not only to allow us to interpret the complexities of the natural world around us, but for the most important survival task of all: dealing with the complexities of our own kind. We have turned these talents to our advantage in a million and one ways of working together and entertaining one other.

What is perhaps worrying is that so many people in business are concerned about the short supply of creative talent they have to call upon. It is talked about as if it were a kind of magic, rather than something ordinary and everyday. We have only ourselves to blame if we kill the creative spirit with bureaucracy and autocracy. Creativity is talked about as a scarce resource, yet every day, in every office and factory, secretaries, salespeople, technicians, and other low-level employees are enacting unnoticed innovations. They are reorganizing filing systems, adding a feature to a promotion, finding a new way to solve some tricky little technical problem. As the saying goes, with every pair of hands you get a free brain.

So what does a business have to do to capitalize on people's creative spirit while avoiding the penalties of irrationality? Here's an eight-point plan:

1. Watch how you manage errors and mistakes. Zero-tolerance cultures drive out exploration and prevent learning.
2. Train your managers to create a climate of psychological freedom in which curiosity is valued.
3. Give people space to express their emotions and time for reflection. By all means set people-challenging goals, but don't hold their reins too tight.
4. In areas of high information flow and complex decisions, don't trust your instincts. Use decision-making aids and statistics before venturing into high-risk areas.
5. Make sure that the climate is one in which diverse expertise and opinions get a real open airing.
6. Question your own assumptions and conventional reasoning before making any important decision.

7. Look out for herding behavior and challenge it with hard evidence about the merits of various options.

8. In decision-making hierarchies, keep a low power gap between the top and bottom; lower-level employees should feel free to challenge the judgments of those in authority.

· 6 ·

Gene Politics: Family, Friends, and the Company of Strangers

GENE POLITICS is how kinship—both real and analogous—affects people at work. Politics can be defined as the art of managing diverse interests to achieve one's goals. In this chapter we consider how evolution has channeled our ways of thinking and feeling to do this in relation to family, friends, and strangers.

THE NEW SOCIAL ORDER OF WORK

WE HUMANS are the most social of species, and for good reason. We are puny, furless animals who can barely survive and can accomplish little without one another's help. It is natural, therefore, that our mental powers come most into their own when dealing with people. Our memories, intuitions, expressive talents, and the flashy skills we like to show off, all are designed to help us get along in a world where most harm and good come from our own kind. For the most part we get along pretty well, and in fact it is imperative that we do so. Our realization of this shows in the high esteem in which we hold individuals who contribute the most to social cohesion and collective achievement.

But some fundamentals of social existence have altered from

those of our ancestral world, and the differences have created some new difficulties for us.

◇   In the past, our daily existence involved almost continuous face-to-face interaction.

It is now possible to live a life of minimal social interaction by dealing almost exclusively with disembodied information or machines. The Internet and other modern media increase this tendency. The depersonalization of social life begets isolated indi viduals who feel a lack of recognition and responsibility—like the hackers who vent their frustrations on the world by unleashing computer viruses on it. Tellingly, the fiendish 1999 "Melissa" virus was written by an aggrieved man in memory of a failed relationship.

◇   In the past, the community within which we evolved consisted of people to whom we were bound by ties of kinship—an interlocking community of families in which we had at least a nodding acquaintance with everyone else.

Today our world is full of strangers. This means that we often get wrong-footed in our social relations. We find we have made a false assumption about a person, behaved inappropriately, been cheated, or gotten into a needless fight. The whole industry of customer relations is designed to counteract and neutralize the negative effects of remote interactions.

◇   In the past, work and play took place in the same common spaces, in the company of a mix of immediate and distant relatives.

Now, in many parts of the world, home and work are separate domains—kin in one, nonkin in the other. The separation of the worlds of family and career places strains upon relationships and creates conflicts in people's minds. Stress spills over from one domain to the other. Many mental health problems and marriage breakdowns are caused by this compartmentalization.

◇ In the past, we dealt with the full diversity of human types in the relatively closed community of our clans.

Now when we go to work we engage selectively with people who are like us in one or more ways: skills, values, education, background. The cloning of work cultures and subcultures generates petty tribalism within functions, departments, and hierarchical levels. Newcomers to a business are often struck by the cliquishness they see in the groups that congregate in the canteen at lunchtime or in bars after work. After a little while they become part of a clique themselves and cease to be aware that the cliques exist.

No doubt there were cliques in the world of our ancestors, but there was one vital difference: degree of relatedness would have been an ever-present consideration in how people dealt with one another. Mostly, gene politics points us in the direction of helping our kinfolk. Before we discuss this, we should consider one way in which we give very positive attention to nonkin: romance.

SEXUAL POLITICS AT WORK

ROMANTIC INVOLVEMENT in the workplace is one of the most common and least discussed phenomena of business life.

In Chapter 2 we considered how evolution engineered distinct but interlocking strategies for choosing mates and allocating responsibilities to men and women. Sexual attraction to nonkin is necessary to expand the gene pool, to avoid the risks of genetic damage through incest, and to optimize our offspring's genetic profile by giving them pieces of two unrelated immune systems. Falling in love is how evolution caps this plan with a commitment-binding emotion. The purpose of this very special emotion—whether between adults or between parents and children—is bonding. Because of our anatomy (large head and a

small pelvic arch—the latter a feature adapted to upright walking), humans have to do much of their growing outside the womb
to reach their full big-brained potential. This means that men
and women need to stick around and help each other, especially
during the early child-rearing years, and to feel the attachment
that will make them nurture their acutely vulnerable dependent
offspring.

The love drive can be switched on at any time, but it mani-
fests itself in both sexes especially during their time of peak fertility. It is estimated that more than half of people's principal sexual
relationships originate in work settings. And the only reason this
figure is not higher is that many men and women continue to
work in sex-segregated environments.

Love is the ultimate kind of favoritism, so it is easy to see
why people take a dim view of it in the workplace. Romance
completely undermines the rational order of the organization. It
fosters partiality and makes people vulnerable.

Press reports in March 1999 recounted how sixty-eight-year-
old Rupert Murdoch, the fearsome boss of the giant media company News International, had suddenly started behaving to his
terrified underlings like an almost warm and cuddly human being,
doing extraordinary things such as smiling at other people. The
reason was that he had fallen in love with thirty-one-year-old Chinese TV executive Wendy Deng, who clearly had ideas about how
her man could make a better impression on others. Love is the
starting point for family, and it is no surprise that Murdoch's son
and daughter by his previous marriage—both prominent executives in the same business—were reported at the time to have
some misgivings about Dad's new engagement.

A sex-segregated environment does remove the intrusion of
heterosexual romance. In hunter-gatherer societies there is generally a mix of unisex activities and tasks and other times when
there is free intermingling and cooperative labor. What seems
most unnatural in modern organizational life are those environ-

ments where a small minority of one sex works in the midst of a large majority of the other. Where the majority is male, as we have seen, the result is often various kinds of oppression and sexual harassment. Paradoxically, it is equal opportunities legislation that has made this kind of disparity common. To avoid these problems, the answer is not to try to shut out the minority sex but to move faster toward equal numerical representation, though this would mean having to put up with the inevitable rash of romances that follows.

THICKER THAN WATER: GENE POLITICS AND THE FAMILY FIRM

GENE POLITICS is the process of how relatedness influences social behavior. It was a breakthrough in evolutionary theory that recognized how self-sacrifice could make sense when it benefited others carrying the same genes as the willing sacrifice. The event is commonplace among humans: many parents have risked their lives for a child.

When it comes to business, in some parts of the world taking advantage of family connections is accepted as normal. In many developed economies we tend to denigrate this as nepotism. It is disapproved of, like love, because it controverts the rational order of bureaucracy. Yet in one major area of organizational life it is accepted and valued. This is the family firm. It is easy to forget how pervasive family firms are. They make up nearly 80 percent of the world's businesses, including many of the biggest companies. Forty percent of Fortune 500 companies are family owned and controlled. Family firms sit halfway between the way of working and living of our ancestral communities and the modern world of the impersonal bureaucracy.

People who work in family firms are often struck by the different "feel" there is within them. There is often a pervasive sense of belonging. Emotion is more freely expressed. The dominant values of the culture are often loyalty and shared commitment. The leaders are dedicated and hang on for the long term because they have a real stake in the business. It is genetic—the firm is part of the inheritance the boss has to pass on. Success comes when the company grows and its assets diffuse across a growing family circle, reaching the point where the equity becomes part of the public domain. Many a giant business has grown this way. IBM started out as a business run by father and son, even though the Watsons never had more than a minority shareholding.

Family firms can also, of course, fail. Only a small number survive beyond one or two generations. Businesses like Johnson Wax, where succession has passed through four generations, are a rarity. There are several reasons for this.

A principal problem is the gene politics of equity division. When the chips are down, I will put myself out more for my brother than my cousin and more for my cousin than a nonrelative. Brothers standing shoulder to shoulder built many leading corporations, like the Walt Disney Company, Marriott International, and Johnson & Johnson. Yet many family businesses are torn apart by sibling rivalry. Gene politics creates divergences as well as shared interests within families.

Two parents, sharing equal proportions of their genes with each of their two children, want what the Talmud calls "peace in the house"—for their two 50 percent shares to get along and help each other. Each child may share 50 percent of their genes with siblings, but each is inclined to put their own 100 percent first when competing with each other for parental resources! Many family businesses fail because parents try to spread their inheritance equally among their offspring.

The patriarch of Smallville Plastics, a family firm, suddenly dies. In his will he bequeaths an equal share of his equity in the firm to each of his sons and daughters. Their portions are equal but not proportional to their drive and ability to run the business. The result is that the most motivated and competent sibling, the daughter, is continually frustrated by the interference of other family members to the point where she sells her interest and quits. The business drifts rudderless for a while and then folds.

Gene politics also splits extended families:

Under its founder, a food-manufacturing firm has become highly profitable with its specialty snack product line. After the founder's retirement, the leadership of the company passes to one of three siblings. This individual lives the life of a playboy, gambles away a substantial chunk of the assets, and then flees abroad in disgrace. His sister, a lawyer by profession, who holds a substantial chunk of the firm's equity, encourages her husband, the finance director of the firm, to mount a rescue operation and take over the running of the business. Her mother and the other brother, both minority stockholders, agree to the plan but soon find they disagree with the direction in which the new boss is taking the firm. This develops into a family feud that paralyzes and eventually destroys the firm.

The genetic interests of parents and children can also diverge in other ways. Parents, wanting the best for their genes, can be dangerously overcontrolling and autocratic toward their maturing children. This is especially a problem in family businesses, since single-mindedness and an inflated ego are often part of the profile that motivated their owner-founders in the first place. It is very common to find a company's founder hanging on too long as head of the firm and then continually interfering from a position of semiretirement. IBM survived only because Tom Watson, Jr., was tough enough to stand up to the tyrannical authority of his father, Tom Watson, Sr.

It is also common for children and parents to have quite divergent ideas about what is best for the children, especially as the latter become independent. This is the territory of Shake-speare's *King Lear,* in which undutiful daughters ally with their husbands—their future genetic interests—rather than with their past—the interests of their foolish and arrogant father—eventually resulting in his destruction.

It is succession that exposes the major weakness of family firms: How can the founder's identity be transcended? Here are some suggestions for owner-bosses thinking about the future of their business:

◇ Set and stick to a succession timetable. Find some enterprise outside the business to commit your energies to.

◇ Widen the leadership succession pool beyond your immediate family. Don't expect your children to share your leadership qualities.

◇ Don't look for a clone of yourself among either your children or other possible successors. This may be a time for strategic change, and you should leave this to others who have a vision different from yours.

◇ Do not distribute the firm's equity evenly among your children. Find other ways to reward the offspring who will not play a major part in the future running of the business.

BEYOND OUR KIN: THE "AS IF" PRINCIPLE AND PSEUDOFAMILY RELATIONSHIPS AT WORK

RATS ARE very sociable and loyal creatures—within their kinship group. Toward nonkin they can be extremely hostile. They are equipped with an unerring sensory perception of genetic relatedness. One brown rat can tell for sure by a single sniff whether or not another rat is a member of its extended kinship group.

We humans lack this ability, which is just as well. We need to cohere in fairly large communities to get maximum advantage from one another so ambiguity about our relatedness is no bad thing. It is in our genetic interest to be sociable and cooperative beyond our kinship group. Evolution has traded on our inability to see our genes or anyone else's by making us generally friendly. However, we are still programmed to be discriminatory in various ways—what I shall call the "as if" principle.

This is the idea that we act "as if" we could spot genetic relatedness. It affects us in two ways.

First, it shows in the way we search for similarities and differences between other people and ourselves. Many people become fixated on superficial features, such as other people's physical similarity to themselves. This, of course, is the basis of racism. We are also inclined to favor people who think and act as we do, people of the same class and culture. Our bias goes further, to the level of looking for and doing deals with those who share our interests and opinions. This general attuning to similarity is clearly something that, in the long run, tends to support our own extended kinship group. It makes sense in the absence of any more objective basis for discriminating in favor of those who share our genes.

The second way in which the "as if" principle operates is that we enact the gene politics of kinship with people who are nonkin. Many a time we treat bosses as if they were parents, subordinates as if they were children, and peers as if they were siblings. Why do we do this? The gene politics of families are hardwired in us as biases toward *presumed,* not actual, kin, since, unlike the brown rat, we do not have any ability to spot shared genes. Such a general bias cannot be reliably switched on and off, so it infects all our relationships.

When the two parts of the "as if" principle act together, they result in such phenomena as groups of men forming gangs in which they treat one another as brothers and women modeling their relationships on that of sisters, especially in the spheres of religion and political activism. Bosses talk about their firm "as one big family," though many firms, in fact, operate like dysfunctional families: bosses act like autocratic parents; staff behave like surly, rebellious children; people do favors for their friends as if they were part of an extended family.

In developed economies we brand this as corruption, yet throughout much of the world it is commonplace. It is regarded as acceptable for people to give advantages to friends and their relatives in hiring, awarding of contracts, and granting access to information. Even in such a bastion of bureaucracy as the European Union this kind of favor has been rife—in 1999 the entire European Commission was forced to resign because of rampant nepotism, cronyism, and self-enrichment! In business the same forces are at work when we use networks to give and receive introductions and advice, but we don't like to think of them as being in the same category.

The "as if" logic of family relations is perhaps most evident in leadership succession, where a departing boss acts as if the leadership were his inheritance to bestow on a favored child. The corporate politics of a family saga may be acted out even when there is no real family. What happened in British retail house Marks & Spencer in 1998–99 is typical.

For eleven years Sir Richard Greenbury was chairman and chief executive of Marks & Spencer, one of Britain's most successful retailers and most admired companies. Built up over 115 years from a Jewish immigrant's street-hawking barrow, it became a byword for quality clothing at an affordable price, much patronized by the British middle-class shopper. It enjoyed profit margins that were the envy of all its competitors, in recent decades cautiously diversifying first into food and then into financial services. It moved with equal caution into overseas markets.

The firm and its stores are like an old-fashioned British family: slightly authoritarian, conservative, and caring. Staff receive a variety of medical and other benefits not normally provided by U.K. retail firms.

Leadership succession has always been a slow, smoothly enacted process. The top man has almost always been a company person born and bred. Some have worked their way up to the chairman's office from lowly store positions. Although M&S is now a public company, it has resembled in many respects the large family firms of the Far East. The chairman is an all-powerful father figure. In the M&S culture it is expected that he will drop in on stores and meet staff—and he does. New product lines and other operational details do not escape the chairman's scrutiny.

After thirty-five years in the company and eleven years as its leader, more and more people were wondering about when Greenbury would retire. He had already declared that he would carry on after his agreed-upon retirement age of sixty to sixty-five, and time was moving on. People were talking about who would take over. Like many longtime leaders he seemed reluctant to let go, and the choice of successor was continually put to one side— at least until the backwash from the 1998 Far East market crisis turned an indifferent M&S company performance into a poor one and the board of directors started to press for a decision on suc- cession. At this time Greenbury indicated he favored his protégé Peter Salsbury, another M&S lifer. Normally, that would have been that. But one man wasn't going to take it lying down. This was Greenbury's dynamic and charismatic deputy chairman, for- mer finance director Keith Oates. To the outside world Oates was the natural choice—a man with a reforming vision. To insiders Salsbury was the conservative choice—the classic M&S retailer. Oates's allies in the company started to wage a war through the media, pressing his claim to the job. Oates himself put forward to the board a bold and attractive plan for a shakeout at M&S. Both the man and the plan had great attractions. Insider support seemed to be split. After one damaging press report about this infighting, Greenbury was forced to fly back early from a vacation in India. He found the company in crisis. Immediately he rallied support, called an unscheduled board meeting, and brought the final showdown to a bloody end. Oates was forced into retirement, Greenbury clung on as nonexecutive chairman, and Salsbury took over as CEO. The whole affair was a serious humiliation and created major embarrassment in the M&S culture. A little while later, in 1999, Greenbury, confronted by the further ignominy of falling sales and shareholder value, stepped down completely. In 2000, against a background of continuing poor performance, the board broke with its "as if" family tradition, bringing in a complete outsider, Belgian retailer Luc Vandevelde, to take over as chairman.

Here we see the kind of Shakespearean politics that commonly exist in family firms:

1. Dad (or Mom) is reluctant to let go of power.
2. Dad (or Mom) expects the children to be faithful, peaceful, and cooperative while they wait to see which of them will get preferment.
3. The children don't see it that way; they lose patience with Dad and Mom and fight with each other to get the favored succession spot, sometimes figuratively assassinating the parents in the process.

The lessons of the "as if" principle parallel those I have suggested for the family firm. The process starts with letting go—a shift in the leader's mind-set from patriarchy to stewardship. A firm is not an estate to be bequeathed but a responsibility to share and shed. Because the "as if" instinct is a metaphor, we can think our way out of it. If we see what we are doing, we can shift from acting out parent-sibling-child relations in our working relationships and become more effective adults.

Another common "as if" pattern is for incoming leaders to remove top-placed insiders and replace them with their own henchmen and -women, like the lion whose first act taking over the pride is to kill the cubs of his predecessor. Leaders cement their positions by an equivalent restructuring of the local gene pool.

PLAY AND THE POLITICS
OF FRIENDSHIP

THOUGH FAMILY relationships are important and formative, a growing child soon becomes aware of a new critical source of learning and experience: playmates. Evolutionary psychologists have observed that the rough-and-tumble fights of boys follow patterns of role rotation that are practice models of the fights our ancestors would have had to act out as adults. Girls play at adult maternal and nurturing roles, as well as the arts of diplomacy and self-presentation. In middle childhood—up to puberty—boys and girls spend most of their time with their own sex, engaged in play-cum-experimental-rehearsals of later-life roles and relationships. These rehearsals are essential to future cooperation and competition. Many a business owes its existence to this kind of play. In fact, one could say that the whole of Silicon Valley stems from gangs of young men who carried on playing together beyond their college years.

On one side of the Valley were Bill Gates and Paul Allen, a couple of brilliant college dropouts who became obsessed with the emerging electronic technology that was to become the PC. Gates was more the driven man and the dominant force, even though he was two years Allen's junior. Gates was (and remains) obsessed with winning and swept into business history, leaving Allen a wealthy but largely sidelined figure. When it came to selling to IBM the idea of the operating system that would make him and his pals multimillionaires within a few years, he enlisted the help of another college buddy, Steve Ballmer. Ballmer, elevated to chairman of Microsoft in early 2000, had less of a nerdish image than Gates and more hard-nosed business sense. He was critical to the sales pitch that launched what would become the planet's most powerful corporation.

On the other side of the Valley was the visionary behind the PC, Steve Jobs. Jobs came from the 1970s hippie subculture of California, where he spotted and exploited the high-tech genius of college dropout Steve Wozniak, an electronics inventor. Lacking the Woz's talent, he made up for it in vision. The Woz just wanted to play with his toys but, under the spell of Jobs's enthusiasm plus a measure of nagging and bribery, came up with the brilliant innovations that would become embodied in the legendary Apple II computer. Jobs's thrusting ambition swept all before him as he gathered teams of buddies to help him realize his ambition—an antiauthoritarian crusade against the corporate establishment might of IBM.

Male buddydom is probably the single most common platform for the origination of business. In almost every business start-up, a founder has used friends to help him get going. Many of the world's most successful companies retain the name of the friendship that founded them. Hewlett and Packard, Procter and Gamble, Marks and Spencer—the list of businesses built of partnerships is very long indeed. In many cases it was the strength of the friendship rather than a business vision that drove the partners toward success. Bill Hewlett and Dave Packard fooled around with a lot of ideas before hitting on the plan that eventually won them success. Masaru Ibuka and Akio Morita's building of the Sony Corporation was a similar story of determined trial and error.

Many of today's firms were founded by the playful enthusiasm of engineers. So it was that Bill Harley and Arthur Davidson's infatuation with mechanized power and speed led them to build and sell a single motorcycle in 1903; within a few years they were selling their legendary bike to thousands worldwide. Other kinds of complementarity slowly built great empires. It took two decades for William Procter and James Gamble, candle and soap maker respectively, to get a factory up and running. What brought the business to its subsequent phenomenal success was the close alignment of their philosophy and values. A work ethic of thrift, hard work, responsibility, security, and co-ownership enabled them to achieve a competition-beating tightness of focus and customer service.

The secret of success in buddy business ventures is trust plus complementary assets. Trust requires shared values and vision. Complementary assets means skills and personality. It almost invariably requires that one buddy allow the other to be ascendant and that the dominant partner possess the leadership capacities we looked at earlier. One effect of this asymmetry is that it is common for an initially useful junior partner to be jettisoned quite soon after the business is up and running.

Today we are seeing an explosive growth of e-commerce businesses, most of them founded on buddy partnerships. Which will survive and which fail? The advice we can give them, from an EP take on business history, is to do some cool self-appraisal on the following issues:

◇ What is the basis of your trust in each other? Are your values aligned closely enough to endure the growth and change of your business? Are your personality and style complementary enough to sustain changes in your relationship, which will inevitably occur as the business grows?

◇ How complementary are your talents? Do your skill bases have different half-lives, so that one partner's contribution will decline as the other's grows? How will you handle this?

◇ Who is the leader? Does either of you really have what it takes to run a growing business? If not, are you prepared to bring in someone who can?

A final thought on buddydom: There is, to my knowledge, no major corporation that was founded by two or more women. Clearly the same male biases that operate for leadership generally also apply here. Men, especially young ones, are more interested in the aggressive play, obsessive venturing, and intensive risk taking that stimulate many a new business. Also historically, there have been powerful male biases in the technical domains where many mighty firms planted their roots.

The e-commerce revolution will undoubtedly see, for the first time, a significant growth of female-owned businesses. We can predict that they will be successful to the degree that they capitalize on the special gifts of women for networking and cooperation, but they will face two threats. One is that as their success grows, dominant males will seek to take them over. Two is that if they grow into conventional hierarchical businesses, men more than women will be motivated to do the politics, design the systems, and operate the controls that keep them running.

TRADING TRUST:
THE ENLIGHTENED ALTRUIST

PRIMITIVE CRITIQUES of Darwinism made much misleading use of tooth-and-claw imagery. Our genes may be selfish, but they have to program us to be unselfish enough to cooperate with nonkin as well as kinfolk. Our survival, which is based on the existence of the tribe, depends on it. Unselfishness comes about quite easily in the communal world of a clan of fairly fixed membership, where birth and death are the main entry and exit routes. All members of such a clan know all the other members and share a common fate with them. It makes a lot of sense for strong norms of sharing to be developed.

Food sharing is the prototype of altruism. After a good day's hunting, what are you going to do with the surplus after you and your kin have had their fill, for meat cannot be stored for long? The answer is to share it, especially among those you value as friends. The next time you've been out hunting and come home empty-handed, you stand a good chance of getting a return from someone you shared with in the past. In this way sharing evens out a group's food supply. It is a naturalistic economy that reduces the uncertainty of the supply of resources. We know that the instinct to share is not the product of intellectual reasoning, as species without our brains or language gifts do the same: vampire bats, for example, share their surplus of cattle blood with nonkin who have had a lean night's hunting.

Working life is full of instances of selfless generosity:

◇ A sales rep gives unsecured credit to a customer.
◇ A supplier turns down a valuable new contract because it would mean reneging on a less valuable but more established relationship.
◇ A teacher covers for a colleague who needs to get out of school early to attend to some urgent private business.

In each case we can explain away these decisions as having an "ulterior motive," but that would be missing the point. People everywhere, all the time, do unnecessary kindnesses just to keep their business going and to make someone else happy. The highly integrated value chain that made Japanese business the most powerful in the world was built on a framework of trust underwritten by a network of unwritten contracts, friendship-based financing deals, and informal supplier relationships.

Trust is probably the single most important value in business. It is what holds cooperative alliances together, enables difficult and dangerous enterprises to be pursued, and gives people the strength and confidence to take a long-term perspective.

The basis of such cooperation is often the possibility of delayed return, the belief that one's altruism will be reciprocated in the future, that when you are down on your luck you will be able to call upon a person you helped. This is certainly true of many deals. They are pure examples of enlightened self-interest. But it goes further than that. There are many situations in which people give without any expectation of return.

It is your last day working for Mega Corp. before you emigrate to a country on the other side of the globe. There is no chance you will see or hear from any of your work colleagues again. You decide to go on a tour of the building. Your first call is on your good buddy who shared your office. You say, "I've always liked that paperweight you have on your desk. I'd really like it as a souvenir of our time together." She gives it to you, for sure (though some embarrassment may accompany this exchange). Next call: your boss. "Listen, I really need to have an open letter of recommendation that I can use for future job searches. Could you write one for me?" This will take your boss some time and effort, but naturally she agrees. You continue on your way. Your tour of the building yields you a bountiful haul of goods and favors, and for none of your donors is there any conceivable future return.

The conventional explanation of such beneficence is that even though the giver knows it won't be reciprocated there is nonetheless a valuable payoff—a warm glow of moral satisfaction. This comes partly from your benefactor feeling happy in the knowledge that you think well of her, even if fleetingly. We are certainly hardwired to care about what others think of us and to try to maintain our reputation as worthy people. But look at some of the things people do, and you will see that it goes a lot further than this.

People give freely to others not only where there is no hope of return but also when no one is watching to applaud or approve of their generosity. Panhandlers survive on the streets because of this impulse. People's lost wallets are returned without a penny removed. Charities benefit from large anonymous donations. And we can't always assume that the people who do these things are even getting the "warm glow" benefit. Many donors will say they are acting out of moral principle—simply following their conscience.

This points to the deeper origins of altruism. It is a mental module that exists around the importance of self-worth. In a world of deal making with nonkin communities where everyone has at least a nodding acquaintance with everyone else, reputation counts for a lot. One could get rich, maybe, by occasional covert cheating. Some people certainly do that. But for most people it is too much effort and far too risky. The safest game to play is to try always to act reliably so that you will require a public reputation for being trustworthy. The best way to do this is to have a consistent image of yourself as a moral person. With this in your head you don't have to go around being a method actor, having to worry at every juncture about whether you are acting "in role." It is simpler and more effective to have your morality deeply internalized and unconscious. Then you can go around acting like a moral person even when no one but you can see what you are doing.

Religious doctrines center around this idea. Some cruder faiths rely on loss avoidance and delayed payoffs (heavenly choirs versus fire and brimstone), but the most compelling creeds draw upon ideas of developing a moral identity via the exercise of conscience. Guilt and shame are psychological shortcuts to avoiding the risk of being found out and punished. This programming keeps human society and most of our business afloat through moral feelings: of debt, obligation, rights, trust, honor, and fairness.

Altruism, although a fact of human existence, is extremely variable. People are born with and early in life develop different susceptibilities for morality. On the whole, women are more generous and kind than men, but there is a wide variation within both sexes. The desire to nurture others is a basic dimension of personality, albeit one of the more conditionable. The way we were raised is important to activating our adult altruistic impulses. Brutal bosses were usually treated brutally themselves in childhood. Men who grew up with many siblings, especially with sisters, become more nurturing and caring than only children or men from small families. This suggests that the conditions of the clan's natural kindergarten—lots of siblings and cousins together—would optimize niceness among men. It also implies that the modern trend toward smaller families has played a part in the creation of today's generation of tough-minded businessmen. The extended families of developing economies such as those of India, the Middle East, and Africa are linked with the higher value those cultures place upon nurturance as a managerial virtue.

The chief implication of what we know about altruism, though, can hardly be that we should all increase the size of our families. It does suggest how important altruism is in business:

◇ To create a community spirit in a firm through cohesiveness and stability of membership
◇ To make behavior publicly visible and reward individuals and groups whose actions are manifestly in the interests of the commonwealth

CONTRACT SENSITIVITY AND THE ART OF SELF-DECEPTION

JUST SUPPOSE you lived in a world where there were no strangers. By that I mean that when you walked down the street you knew everyone by sight and many by reputation. Actually, in some areas of your life you already live in this world. In your neighborhood you may know all the storekeepers and all the locals. You are probably indifferent to those who are strangers, the occasional tourists who pass through your village.

Every day, when you go to work, you meet and greet people you know. Again, the few strangers—perhaps the sales reps who come in to meet your colleagues—are irrelevant. You don't have to talk to or deal with them. Even in the heart of the urban jungle strangers are just part of the scenery. Most of our business, both personal and occupational, is conducted with people we know. This suits us. It is what we were designed for. Our ancestral model is one of a relatively isolated community—a conglomeration of families within which everyone has at least an indirect kinship tie with everyone else. Living in such a community has a number of implications.

◇ You have limited choice in your interactions.

◇ You might be able to avoid people you hate, but not completely. It is better not to hate them.

◇ You had better be careful about whom you ally yourself with. It is more difficult to walk away from a relationship than to start one.

◇ It is also difficult to walk away from your own actions. People bear witness to them, and thus they become embedded in the community's collective memory. If you let someone down, not only will that person remember and be likely to deal differently with you in the future, but others will also get to know what happened and your reputation will suffer.

The cost of losing your reputation can be high. People may exclude you from opportunities to share in what is best in the community, be it food or friendship. In hunter-gatherer communities anthropologists have recorded how minimal punishment needs to be to deter transgressors. Social disapproval is a powerful deterrent, especially if it is demonstrated in a formal public arena. For all but the most extreme crimes temporary social exclusion is sufficient. Banishment from the community is the ultimate sanction. In business being fired is the worst punishment. Setting aside the financial cost, the strength of a corporate culture can be measured by how much of a stigma it is to be dismissed.

We are incredibly sensitive to perceived violations of our community's implicit contracts. Look at how angry people get when they think a peer has gotten some undeserved advantage, be it ever so trivial, such as getting a privileged parking spot or jumping the queue for technical assistance. But what happens when one person thinks another has played unfairly, gotten away with something, or broken the unwritten rules in some way? As in road rage (which is also often caused by perceived violations of an implicit contract), *both* parties get angry. To each, the other guy is in the wrong.

The altruism instinct described earlier has a condition attached to it: self-deception. It is essential that we think well of ourselves. Self-deception gives us the benefit of the doubt in the interest of maintaining a consistent positive self-image as a reputable, trustworthy member of the community.

The last time you got into a fight with a colleague, the chances are that

1.   It was about a violation of what one or the other of you believed was an implicit norm or agreement.

2. Each of you was absolutely convinced the other was in the wrong.

The areas of business in which such conflicts arise are those where there is no objective truth and people have a big psychological investment. All too often these conditions apply in the area of performance measurement. Any negative judgment a boss makes of a subordinate (or vice versa in 360° appraisal systems) invites outrage in direct proportion to (1) the subjectivity of the judgment and (2) how central the criticized quality is to the subordinate's self-image.

When Mike tells Marie that he rates her low in interpersonal sensitivity, the effect is far worse than when he points out that she has made errors in purchasing orders she has been processing.

But objective data are not always relevant. Sometimes, as with 360° appraisals, people need to hear what other people think of them. In most businesses the sense of community is not strong enough for this to happen in an easy, natural way; it requires skilled mediation.

Even in the most open communities, it is still difficult. We are hardwired to analyze one another's motives and empathize when we are on the same wavelength but not to see the other person's point of view when it differs from ours. Being an effective social performer requires us to think most of the time that we are in the right.

To counter this requires the very difficult skill of *decentering*. This runs against the grain of our nature. We are self-centered around our self-image and purposes. It is contrary to our instincts to put ourselves in the mind's eye of someone who is seeing things quite differently from us. There is also another big drawback to doing this: it means facing up to the possibility that I am deceiving myself. This is a tough proposition that most of us cannot face up to without support—tough but not impossible. We have to see the benefits, and for this to happen may require the help of a trusted friend or counselor.

TRADING WITH STRANGERS

Now let us introduce strangers into this world. Someone from an unknown tribe walks into your camp. What should you do? You and your friends could kill him. But he is alone and unarmed. There is no tangible gain and a high potential risk. His friends might come after you. So what do you do? Smile. He smiles back. Before long you are trading. It's the best thing to do with strangers.

The human predilection for trading can be seen in every schoolyard. Kids will swap just about anything, from objects they have found to bits of gossip. Trading is older than any other human economic activity and has shaped the entire march of civilization. Nowhere is this instinct portrayed more starkly than in Primo Levi's moving eyewitness testament of his time in the Auschwitz concentration camp, *If This Is a Man*. As the title suggests, the camp regime has reduced humanity to its core fundamentals. One of these is trading. Levi describes the "market" of the exercise yard, where inmates desperately deal in objects of pitiful worth: scraps of metal, bits of rag, and such. The activity is conducted as if any of these things might be of material advantage to survival, even though the likelihood of this is remote. Trading requires little justification, and people do it obsessively for no other reason than because it is something to do. This is well known in the world of finance, where purposeless "noise trading" is a familiar and economically inefficient activity.

But what exactly is a stranger? Most of the new people we come across are fundamentally unsurprising to us. Our "as if" psychology locks in immediately to locate the person in a framework of the familiar, for example, your private community of known friends, enemies, and associates. It is the ability to switch this program on that enables the following kinds of things to happen:

◇ A sales rep calls on a client and finds herself dealing with someone new. She quickly establishes a friendly rapport and conducts her business as usual.

◇ A new associate starts at a law firm. People show him where the coffee machine is, how to fill out his expense forms, and other things that will make him feel comfortable and accepted into the group.

◇ An engineer is sent by his firm to a trade fair to look at new equipment. On the first day he tours the stands, makes a lot of contacts, and ends up at sunset in the bar swapping stories with a bunch of people from a rival firm.

How does such instinctive cooperation work? In the animal world, and especially among primates, there are rules of engagement. The best way to deal with a stranger is on the "as if" principle—the assumption that you are both from the same tribe. There is more to gain from cooperation than conflict, which may result in significant losses. As we have seen, we are powerfully motivated to avoid loss. We are also powerfully motivated to avoid being suckers. What if we cooperate with a stranger or even with someone we know, and they cheat us?

Outright cheating is dangerous because if you are caught the penalties can be heavy. If you have great wealth and power, you are cushioned but not immunized against these costs. In the environment of our hunter-gatherer ancestors no one became really wealthy because of the perishability of food and the equalizing ebb and flow of advantage and opportunity. Most of the time in our closed communities it paid to be honest and cooperative. Only extreme deviants—and yes, all communities do contain them—deliberately cheat for advantage.

Things are different in our world of strangers. Here it is much easier to cheat without being detected. For the sake of a clear conscience, most people prefer to be honest, but the further we depart from a sense of community, the more crime increases. Crime tempts people with the strongest desires, the least to lose, and the most to gain—those who have lost out in the past or feel stuck near the bottom of an anonymous social heap. People such as Nick Leeson, whom we looked at in the last chapter, are motivated to short-circuit the slow and difficult ascent of the greasy pole to success. They play fast and loose with the rules and take big risks. Some of our most crucial mental models are designed for survival in a world where there is a risk that people will act deviantly. These include our ability to conceal our feelings when we interact and our highly attuned sensitivity to cues that suggest that others might be cheating us. Some people—those who become sales executives, actors, and even con artists—have much better skills in this area than others do.

Evolutionary biologists and economists model these dilemmas through game theory. In the classic experiment there are different payoffs for cooperating and cheating. A business example would be when you and a competitor are successfully selling a specialist type of insurance in a growing market. You and your rival confront the same choices: (A) advertise in such a way as to emphasize the benefits of the product and thus increase the market for you both (cooperate) or (B) denigrate your rival to steal their market share (cheat). If you both opt for A, you will both make modest gains. If you both opt for B, you will both see your sales decline because the public will lose faith in both of your businesses and your products. If one of you opts for A and the other B, the cheat will be the big winner, the cooperator the big loser. In experiments, as in real life, the payoffs can be configured so that the rational strategy is to cheat. But people still cooperate. To economists, this is puzzling. To evolutionary psychologists, it makes sense.

In a series of computer-simulated experiments, different strategies for playing this kind of game were pitted against one another. An evolutionary contest was engineered so that programs racking up the most points from payoffs were allowed to "breed" more copies of themselves. Eminent mathematicians, biologists, and others submitted a variety of ingenious and complex strategies to compete in this Darwinian contest. Throughout the various trials and models the great surprise was that one program of remarkable simplicity kept emerging as the consistent winner: "tit for tat." This instructs you to cooperate on your first encounter and thereafter imitate whatever your opponent did last time.

It is this program that tells us to smile at strangers and then see what they do before deciding whether to go on being nice. It is prudent cooperation, kindness with justice, warmth with toughness. It is treating strangers initially "as if" they were part of our community.

In a closed community, one can expect people on the whole to cooperate, because (1) you never know when they might have a chance to retaliate and (2) you want to keep your "nice guy" reputation intact. Companies producing similar products usually desist from attacking one another directly for this reason. To live and let live is easier. Also, in a growing market why waste your energies fighting others?

Generally we prefer to cooperate and share. Psychologists have done experiments that show that most people will turn down opportunities to cheat strangers for personal gain, even when there is no chance of retribution. It is a different matter once negative feelings engage.

TOOTH AND CLAW:
AGGRESSION AT WORK

NATURALLY FRIENDLY creatures though we be, we are capable of terrible cruelty and barbarism. Some of the worst acts in business are carried out in cold blood, such as environmental pollution, mass layoffs, and deals that put other companies out of business. Mostly these are simple self-interested behaviors enacted impersonally. It would be a mistake to see them as examples of aggression.

So let us concentrate on the hot-blooded variety. Human aggression is aroused by the same causes as that of many other species. These broadly fall into four categories:

Defense Against Threat

Even normally passive creatures will turn on all the aggression they possess if they see they are cornered and threatened with imminent loss.

> Iris hears the news that her division of Mega Corp. is going to close, resulting in mass layoffs. She realizes there will be no escape for someone in her position. She is angered at the unfairness of it all. She walks into her boss's office in explosive rage. After delivering the boss a piece of her mind, she storms out to cool off over a long lunch.

Competitive Contest

Members of the same sex of many species will often engage in aggressive competitive displays to assert their superiority over one another. There doesn't have to be any objective other than demonstration.

Fred and Harry seem to have rows every time they are in meetings together; they seem determined to take different positions on almost any issue that comes up. There is nothing tangible, like promotion, for which they are rivals. It is hard to see what either of them can gain from their fighting.

Assertion of Position/Identity

In many social species an individual who is unsure of his or her status or has been put down by another of higher status may react by picking on a weaker individual who can't retaliate.

Ever since Scott applied for and failed to get a promotion that would have moved him to head office, he seems to be out to get Alec, the newest recruit to his staff team. Alec is on the whole an efficient and hard worker. Several times Scott has publicly criticized him in a way that is plainly painful for Alec and embarrassing for everyone else. Everyone feels that Alec is being unfairly picked on.

Kinship and Group Protection

Almost all creatures will fight fiercely to defend their offspring. Some display aggression against any strangers from outside their extended kinship group.

Katie is the manager of a bank branch office that has just merged with a former rival institution. It is her job to oversee the integration of the two staff groups. After a complaint by one of her staff she conducts an inquiry and is shocked to uncover a catalogue of backbiting, misdirection, rumormongering, and protectionism operating between the groups.

Aggression is always self-centered. The heat of feelings focuses on possible loss, contract violation, desire to achieve, or perceived differences in style or substance. The tendency for conflicts to escalate makes decentering—assuming another's perspective—out of the question as a solution, especially when commitments to hostility become public. It would take some intensive group therapy at a location away from work to get Fred and Harry in the second vignette, above, back on good terms. Even then their personalities might prevent a final resolution of their ongoing conflict.

People often brand conflicts as being due to "personalities." This is almost always mistaken. It is true that some people differ so much in style that it is difficult for them to get along together. This is more a barrier to communication and cooperation than a cause of conflict. Some reason, however minor, is needed to exploit such a difference and turn it into open conflict. In business the causes often lie in the arrangement of rules, roles, and responsibilities, which set people against one another.

One of the most pathologically divided organizations I ever saw was a police force. Like many militaristic organizations, it had a single linear hierarchy for advancement, with a competitive system (a promotion board) to regulate the upward movement of individuals from one level to the next. It was never explicitly stated what behaviors officers would need to demonstrate to be rated successes in these competitions; they were simply given a wish list of ambiguous adjectives such as "leadership" and "integrity." The interview part of the process was conducted by officers with whom the candidate normally had no operational contact. The outcome was a simple pass-or-fail decision, in writing, after a long delay and with no explanation of the reasons for the decision.

The consequences were far-reaching. Junior officers expressed deep hostility and mistrust toward senior officers. There was widespread superstition about what you needed to do to get promoted. The culture was highly loss-averse: officers believed that achievements mattered less than having a clean sheet without any blemish. One conspicuous failure would wreck your career chances. This created two groups of officers: one group engaged in aggressive peer competition, while the other had concluded that they were no-hopers and had given up trying to get promoted. They rationalized that the lower-status jobs were the best, while expressing contempt for their superiors. For both groups criminals were a focus for displaced acts of aggression.

In this case the macho culture gave a very sharp edge to the conflict, but the causes lay in the way the department's working arrangements and systems set people against one another. They do so in many organizations. Here's the recipe for doing this:

◇ Make status and achievement scarce resources.
◇ Make failure an ever-present threat.
◇ Force competitors to share limited resources with one another.

This recipe will generally cook you a hotter set of relationships if the competitors are men. Some primate males beat their chests in victory, and one of the most unattractive features of human aggression is the euphoric exultancy of male winners over losers. There are disturbing reports of how men get high on killing in warfare, and every day, in some workplace or other, some man is rejoicing over a victory over an enemy or rival. Marketing "warfare" and corporate "raiding" bring out these sentiments without difficulty.

From what dark well do these feelings spring? For men, they are a raw affirmation of essential power—a deep reassurance as to one's ability to avoid loss and to triumph in competition with others. Sports give license to similar sensations. We clearly like to ritualize conflict in sports and entertainment. We play "as if" games in which we (especially men) overcome others by force (including force of argument). We (again, mainly men) are attracted by the vicarious pleasure of watching other people have public fights. In business and academia we have a favorite arena for such conflicts: the conference.

If you go to a conference within your own culture where you don't know anyone, you will mainly be ignored, especially if you are young (and not an attractive female). One or two people may sniff around to find out if you have any status or tradable information of value, as if you were a stranger who had wandered into camp. To get people at such an event to attack you would be quite difficult. It would take an act of direct challenge. For instance, you could stand up and attack a speech you have just heard. If the speaker is eminent, he is likely to ignore or disdain you. By acknowledging you as a threat, he would run the risk of lowering his own status. You are much more likely to provoke a counterattack from some young hopeful.

But the bloodiest contests occur between people who know each other. You will find it easier to get into a fight if you come with a reputation; that is, you are not a real stranger. If you are eminent, some young gunslingers may want to take their chances with you. It raises their status with their peers if they can skirmish successfully with someone of high status.

Biochemistry is the key to aggression. It is no accident that evolution times the maximum flow of testosterone in the bloodstream peaks to coincide with the early years of male adulthood, when the contest for status and mating opportunities is likely to be sharpest. But as physically weak and vulnerable creatures, humans prefer, if possible, to compete on criteria that offer more lasting advantage, such as refined skills, intellectual power, or artistic prowess. But many men, through disadvantages of birth or inheritance, have little hope of rising via these routes. It is for this reason that much of the violence on our streets is perpetrated by members of the underclass. More sports stars come from poor than wealthy backgrounds. More intellectual achievers come from advantaged backgrounds.

But in any social group, some individuals are more aggressive than others. Like many of the other qualities we have discussed, aggression is a heritable dimension of personality and temperament. In just about every office you can spot three kinds of angry people: those who fly off the handle at almost anything; those with long-standing grievances and frustrations; and those who get angry when they're stressed out. Facing these are the people who never seem to get angry about anything. We have different and quite fixed thresholds for negative as well as positive emotions.

TRIBALISM AND WARFARE

THE MOST vicious fights in business are between individuals who see themselves as representatives of some group interest. Psychological identification with one group that is in a dispute with another raises the stakes of interpersonal conflict. As we saw in Chapter 2, people only have to be told that they are part of a group to trigger their primal impulse to see their own group as better than others. The addition of competition and an identifiable marker—such as skin color—is all it takes to create discrimination. Graft onto this a mythological theory about why races differ, and full-blown racist thinking and conflict arise.

Tribalistic labeling and sectarianism are all around us on our streets. They are also present in our factories and offices. Regional offices mistrust the head office. The front office and back office bad-mouth each other routinely. Production and sales are locked into mutual misunderstanding. Everywhere, people in one part of a society portray people in other parts in ways the others would not recognize.

In the business world I hear a lot more lethal talk than see lethal action. Most "killing" is done remotely—for instance, when some traders steal business from others who are anonymous numbers on a screen.

What typically happens in face-to-face conflict is a lot of talking-up prior to the point of engagement. Conversations rich in righteous indignation among Production people about "those damn fools in Marketing" get everyone frothed up. So what happens when the gang from Production actually has a chance to confront Marketing in a group meeting? There is a lot of sniping in the shared "code" that everyone understands, but nothing is said directly enough to look like a head-on challenge to combat. That would be risky, a position from which there would be no line of retreat. After some skirmishing at the monthly meeting, the Production folks go away saying "I think we let those idiots in Marketing know the score. They'll think twice before trying to put one over on us." Meanwhile, the marketing people are putting an almost unrecognizable construction on the same events. Honor is served. Everyone feels fine.

Intertribal conflict is often showy and stylized. It has been discovered that ancient Roman generals did not pile their legions into bloody conflict. Instead, armies faced each other, hurling insults and nonlethal missiles, with the occasional set-piece tournament between leading warriors. In this they resembled gorillas who roar and beat their chests but rarely come to blows with their enemies. Actual attack is so risky as to be self-defeating. The scuffles between gangs of youths are like this, often a lot less dangerous than their posturing histrionics would seem to suggest—until, of course, guns are placed in their hands.

This kind of conflict fulfills a primary group function. The external enemy makes a group feel united; having a clear common purpose makes us feel strong. This effect feeds back to each individual member. All the talk of fighting and winning makes people feel better: stronger, superior in status, and more needed by the group. For people of low status, their group becomes the principal context in which they can feel good about themselves.

Hostility between groups is almost always more pronounced

from lower to higher status than vice versa. The angry gossip I hear in business is almost always by the underprivileged against the privileged. Sales hates Marketing more than Marketing hates Sales. HR hates Finance more than Finance hates HR. Everyone hates Head Office more than Head Office hates anyone else. In fact, conflicts among neighbors tend to be much sharper than the antipathy against more distant tribes. When the HR manager from Company A meets the HR manager from a major competitor, Company B, at a conference, they do not start brawling. On the contrary, they both complain about the Finance people in their respective companies.

A time-honored method of suspending tribalism and uniting divided factions is going to war against a common enemy. Is there another way?

THE DUKE ELLINGTON PRINCIPLE: HOW TO CREATE HARMONY

MANY OF the hot- and cold-blooded fights in business can be avoided. Individual aggression takes place most frequently when the prizes and penalties of winning and losing are greatest and when organizational goals and roles force competitors to squeeze through narrow channels of shared resources and limited opportunities for success. Fights between groups are caused by the way structures separate people according to their skills and special interests. Abolishing tribalism requires getting rid of tribes. As long as we organize around personal attributes, such as qualifications and status, tribalism will persist.

The alternative is to organize around something everyone cares about more than any individual characteristics. In a good jazz orchestra no one cares whether the players are white or black, young or old, male or female, educated or uneducated.

There are only two criteria: Do you love the music, and can you play it? Let me call this the Duke Ellington Principle in honor of the great man's embracing worldview, for it contains one major part of the answer to the problem of discrimination in organizations. What is needed to defeat the various isms is some over-arching goal that is compelling and unites the interests of every member of the band—like making a single uplifting glorious sound.

Families are at their most harmonious when they are united in some cause or other, beyond each member's immediate private concerns—such as the fate of the family business. Buddies like-wise get along better when they share a common purpose to which each can make a distinctive contribution.

Our ancestors' world would not have been conflict-free; there are always relationships and resources to squabble over. But two aspects of their life would have reinforced the primary value of harmony. One is the relatively closed world of the clan as community. When there is no alternative on offer—when you are discontented but cannot easily vote with your feet and walk away to find another home—you are going to work harder to make the community one in which you want to live. The other is the ability to organize freely around tasks and shared goals. This ensures a marriage of people's motivation and skills around a single objec-tive, such as a hunting party, craft-working group, or entertain-ment troupe. These two aspects create an ideal setup for the Duke Ellington Principle to operate in. The only conflicts these arrangements arouse are competition for inclusion and frustration over exclusion. These are healthy insofar as they underline the value and importance of the group. Conflict can be avoided if all members are fully engaged—stretched, even—by the require-ment to perform all the activities needed for the community to survive.

Now look at the modern corporation. People come and go as they please, and many have no sense of identification with or commitment to the company. Groups are statutory; many exist for political as much as operational reasons. There is little self-organizing. People are told what to do and how to do it. Many groups are not really teams. They don't actually *do* anything together, a point that has been made about many so-called top management teams. Often what the group produces or does is not recognized or valued by the firm.

These conditions have several effects. People complain about meetings wasting time and effort. Some fight to gain even small advantages within groups. Others are free riders who extract benefits without making a fair contribution. So where can we see the Duke Ellington Principle in practice? Here are some examples:

◇ Many small firms come close to the ideal we have
   described. Some of the new e-commerce start-ups are
   like hunting parties. Everyone has a distinctive role in
   scouting for opportunities in a changing landscape.

◇ Trading desks in investment banks are often highly
   cohesive, with an intense exchange of information as
   people help each other solve problems in an environment of furious change and complexity. Everyone rallies
   around when one member gets hit by a major difficulty.

◇ Groups of firefighters know and trust each other's skills
   and resources. All are conscious that they are powerless
   as separate individuals and that their effectiveness
   depends on the willingness of each to endure and take
   risks for the sake of colleagues. All also know that their
   common objective is more important than their
   personal concerns.

◇ Project teams in aerospace work have a long-term
   perspective, within which they set demanding staged
   objectives. Members have highly refined complementary
   skills. They continually learn from and adjust to one

another. They know that what they achieve will be of lasting value to the firm.

Like musicians in a jazz orchestra, individuals in all these groups improvise within a disciplined framework. All respect and value each other's distinctive contributions. Feedback is spontaneous and positive. The group spirit is so strong that members have a feeling of losing their egos within it.

We cannot expect every group that exists in business to exhibit this quality. It is likely to be fleeting for even the most dynamic of groups. Nor can we close the membership of our business communities to make them clanlike. But we can keep the size of units small enough to engender a sense of intimacy and shared fate, and we can place greater trust in people to organize creatively around what is important to them and the business.

◆ 7 ◆

The Gossip Factory

WORKING WITH WORDS

THE GENIUS of humans is most striking in speech and understanding. More than anything else, our gift of language is what sets us apart from all other known life-forms. It is what enables us to plan, achieve, reflect, and share our experiences. It is responsible for just about everything that shapes our lives.

So if we are born communicators, then how come we do it so badly in business? Of all the failings of organizations, poor communication is the most common. Time and again, a major problem could have been avoided through effective communication. The irony of poor communication in business is that it is usually due to people (managers, mainly) trying to communicate unnaturally while everyone else is trying to communicate naturally.

The board of Mega Corp. holds a secret crisis meeting to discuss the company's falling stock price and what can be done about it. Before them is a consultant's report that recommends radical restructuring, major cost cutting, and refocusing the product range. One of the executives is put in charge of the change program to implement some but not all of the report's recommendations. The areas of greatest controversy are put aside to be considered later. The change program is a disaster. The managers running it have different ideas about what they should be doing. The company climate sours to an extreme level of mistrust.

205

Inaccurate scare stories start circulating. The most talented and valued skilled specialists start quitting, even from those divisions that will be unaffected by the changes. When it comes to implementing the changes, management encounters massive resistance by and noncooperation from staff.

In a change program good communications are the key to success. They typically fail because of lack of openness, involvement, and clarity, all directly due to the hypersensitivity of executives to risk and possible loss. Executives take refuge in the delusional logic that if the outcome (improved profits) is desirable and the changes are rational and reasonable, once they are properly explained people will accept and implement them. It is at this point that managers lose touch with the reality of the way in which humans are designed to communicate.

The themes that govern human communication and the problems encountered in business are several:

$\diamond$ Our instinctive patterns of communication are not rational, accurate, comprehensive, controlled, or unemotional. What we really like to do is empathize, gossip, network, and tell stories. We have a strong oral preference. We love impromptu talking and listening. Many of the communications people find most satisfying are the least efficient for getting business done.

$\diamond$ Some people are better natural communicators than others, but these are often not the people who are entrusted with key communications. Many business leaders lack competence in this area. The skill people find most difficult is decentering—understanding what it must be like to be on the receiving end of a communication. We easily make the mistake of assuming that other people are like us in what they understand and want to hear.

◇ Communication networks in business are organic. They
flow along the twisting channels of the rumor mill.
These are formed by people's desire to give and receive
information as a form of currency or grooming. This
makes networks political. It is a vain hope for executives to believe that a message poured down from the
top of the organizational tree will cascade down in an
orderly fashion through all the tree's branches to its
roots.

◇ Our psychological capacity to network is limited. Once
a business gets much beyond 150 members, it is difficult
to maintain a unitary communications community. Sub-
groupings assume greater significance.

THE TALKING ANIMAL

EVOLUTION DESIGNED us to talk. During this century scientists
have figured out how it has equipped us with a two-part system. We are born with elaborated speech centers in our brains,
loaded with preprogrammed structures for grammar and syntax,
which as we grow link with memory banks designed to encode
and decode lexical meaning. We come into the world with a
highly sophisticated vocal apparatus. It has the unusual design
feature of a windpipe sharing the digestive tract in the upper tho-
rax, with the junction between the two channels policed by a flap
of retractable tissue. This is a risky arrangement, and people
choke to death every day as a consequence of it. When evolution
designs something with a such high risk, it must be vital to the
survival interest of the creature.

The evolutionary design of language is apparent from earliest
childhood. All babies babble, and all parents chatter back to their
infants. A child has to be isolated from any kind of speech for
many years to prevent it from learning to speak. In normal human

society the rate of acquisition of vocabulary and sentence production in the first ten years of life is the most astonishing feat of
learning. The brain is critically receptive during this period, in a
race to equip the growing human animal with its most essential
tool for survival: the ability to make things happen by remote
control.

What evolution did *not* do was give us an instinct for reading
and writing. It gave us an acute visual system that enables us to
see fine differences between symbols, and it gave us the manual
dexterity to hold and manipulate writing instruments. But that's
as much help as we get. We have to learn the rest of the skill by
painstaking practice. And as every parent and teacher knows, for
each infant it is a long and difficult uphill struggle from a zero
base to proficiency in reading and writing. Many never make it.
Even in the most advanced economies illiteracy is common.
Dyslexia, or word blindness as it is sometimes called, occurs
because the physical elements we need to develop this skill do not
always fall into place. Evolution never got around to modifying
our design to accommodate the evolutionarily recent development of literacy.

Our Stone Age ancestors drew pictures and told stories.
Even in the very recent past of human history, the record was
largely oral. The epic poems of Homer and Ovid were imaginative
stories to be memorized and dramatically recounted. They are full
of rhythmic repetitions to facilitate this. We love to tell and listen
to stories.

Now look at what happens in business:

◇   Analysis of managers' behavior shows a strong
    preference for oral over written communication. They
    do much more of it and do it better than other forms.
    Most of a manager's day is spent in short, unscheduled
    conversations with peers, bosses, and subordinates.

◇ Most managers hate to write. Their reports are often badly expressed, and their more general written communications are poorly constructed and widely misunderstood by their recipients.

◇ Many managers dislike reading. They prefer lists, images, or sounds to text. Only a minority of managers read anything more taxing than newspapers and magazines.

◇ Ask employees what forms of communication they prefer, and they will invariably put at the top of the list informal and face-to-face channels.

◇ Although they don't like it and do it badly, managers fall back on formal written communications for many reasons, the chief one being that it is the only efficient way to reach large numbers of people. Also, they are motivated by the politics of corporate bureaucracy: written communications provide a defensible record and visible proof that they are doing their job.

◇ People listen to and rely on gossip for news about what is really going on in the firm. They like and readily believe a good plausible story in preference to a complex or ambiguous fact.

◇ E-mail is causing a rash of new communications disorders in organizations because people use it as a substitute for what it cannot replace: face-to-face interaction.

Look at the advent of the mobile phone. It started as a rich professional's emergency communications kit. Once the price fell, human social instincts took over. The sound of people gossiping is now everywhere about us on every form of public transport, in restaurants, on the street, even in the schoolyard. Eavesdrop— it is hard not to!—and rarely will you hear some functionally essential or instrumental communication. What you hear most is storytelling, gossip, grooming, rumormongering, and network maintenance.

There is no point in decrying these activities. They are essential activities of the human animal, and the mobile phone has merely provided a fresh way in which they can be exercised.

STORYTELLERS: MANAGING THE FICTION IMPULSE

IF YOU want to give a good after-dinner speech or a successful lecture, tell a story. If you want to be the center of attention in your social circle, learn how to recount dramatic narratives—especially about mutual acquaintances. If you want to be influential and attractive at work, give explanations of what is going on in the firm as if you are reporting on a soap opera of heroes and villains. The best stories are the ones that can't be disproved, which means you can get away with almost any explanation of events that hinges on other people's invisible motives and feelings.

It has been argued that the construction of stories involving leaders is the key to their being perceived as legendary and that stories can be powerful tools for mobilizing corporate culture change. Some companies, such as Xerox in the United States and Escom in South Africa, even use them as training mechanisms, to help employees to envisage and identify with organizational change.

Stories are memorable because they simplify and dramatize. For example:

Alan is talking to Amanda, a colleague of his: "Of course you know that Janice [their boss] really hates Marie [her opposite number in a parallel division] and has been trying to get rid of her. She deliberately fixed the X project so that Marie would be exposed." Amanda nods. Yes, she'd thought as much. Janice had said something to her about Marie's "fool-ish" project.

The truth is that Janice and Marie—both high-profile figures in the firm—don't have much time for each other, largely because they are so different in style and approach. But in neither of their hearts is there really genuine antipathy as much as a general lack of interest in each other. Janice actually had nothing to do with the project Marie was assigned to, which was ill conceived and under resourced from the start. In fact, Janice had mentioned this to various people, including Amanda.

But where's the story in this? The truth is mundane. What really happened is hardly worth talking about. But Alan's version has better news value. We prefer dramatic conspiracies to every-day foul-ups. There is a serious reason for this preference: Given our tendency to loss aversion, the fail-safe option is to have advance warning about possible danger from people who have power, even if it proves to be unnecessary. Scare stories alert us.

The fiction impulse is hardwired (mainly in the left hemi-sphere of the brain, for right-handers). It is vitally important to be able to make sense of things going on around us quickly. Throw a bunch of data at us, and we can quickly find patterns and mean-ing in it. The crazy stories of our dreams are the narratives our brains involuntarily construct out of random thoughts. During sleep the brain has work to do: clearing up, filing, and reprocess-ing memories, impulses, anxieties, and the other detritus of expe-rience. Then consciousness steps in. During the somnambulant state between sleeping and waking the conscious mind dips into the stream of nonsense and cannot help itself doing what it is sup-posed to do when we are fully awake: make narrative "sense" of it as a dream. Stories are favorite ways of making sense of the world because they fill in the vital information about people's motives and intentions, and they also give us ideas about what may hap-pen next.

Here's another example of the fiction impulse:

Al, Valerie, and Brian are interviewing candidates for a sales manager position in their firm. Al is head of the sales division. Valerie is the personnel manager. Brian is the former sales manager, who has just moved into marketing. They have spent the morning interviewing applicants and are now discussing them. They have gotten into a debate about one candidate.

"I'm not sure about his motivation," says Brian. "He didn't give a very convincing account of why he wants to leave his current firm."

"Yes," agrees Al. "In fact, his explanations of several of his job moves didn't add up for me."

"I wonder," says Valerie, "if he has trouble getting on with colleagues. In every case he was in a team, yet none of his references say he was a good team worker."

"That makes sense," agrees Al. "He struck me as a bit of a loner. What do you think, Brian?"

"Hmm, now you come to mention it, that does ring true. Did you notice that he didn't once mention people he's worked with when we asked him about his achievements and satisfaction?"

"OK, we're all agreed, then," says Al. "We want someone who's going to be more of a team player. Let's look at the next case."

You might think this is a very unprofessional way to conduct a selection interview, and you'd be right. But scenarios like this are acted out every day in firms all over the world. In each case a candidate's fate is sealed, but on what basis?

What really happened in the case above was this: The unfortunate applicant left his previous jobs for various reasons—the search for more money in at least one case—but he figures that's not the sort of thing you admit to in an interview; it makes you look mercenary. His main reason for wanting to quit his present job is that he dislikes his current boss, whom he finds erratic and unhelpful. To say that in an interview would also sound bad, and besides, these people might know his boss. As for his colleagues, he always got on well with them, so he didn't feel there was any need to mention them. In their letters of reference, no one had anything exceptional to note about his collegiality, and if they had been asked none of his referees would have identified it as a problem. His rather cool manner during the interview itself was due to nothing more than a mix of nerves and self-consciousness.

There is no coherent or compelling logic that ties together the information the panel has been given. They have to come to a decision and need something to help them. Like the jury in a trial, they need a story that puts all the bits of "evidence" together into a plausible whole that forms a coherent narrative. In the movie *Twelve Angry Men,* the dissenting juror, played by Henry Fonda, spends his time demolishing the narrative that the prosecution has sold to the jury members, and which they are intent on embellishing, about the motives of the young boy who stands accused of murder. In real life, the famous defense lawyer Alan C. Dershowitz has described his task in the courtroom in just these terms: to dismantle the narrative that the prosecution is trying to sell the jury and replace it with a more defensible account. This may involve replacing a version based on human agency—will and choice—with one based on bad luck or coincidence.

In trying to do this, the defense lawyer has to overcome the instinctive human dislike of randomness, happenstance, and accident—which actually make many things happen in the world. Not only do we like a story, but we like a moral drama in which people are the central causes of events. Things happen because of the intentions and competencies of good/clever or bad/stupid people. This instinct infects all our reasoning. Psychologists call it the *fundamental attribution error*—the tendency to see events as caused by motivated people's intentions rather than as random or impersonal events.

Which of the following scenarios are more appealing?

1.   A certain business crashes because of bad decisions made by an incompetent boss.

Or:

2.   It crashes because unrelated and technical problems in key operating systems occurred simultaneously, straining the system beyond its limits.

1.   A high-profile manager leaves a firm because of a plot against her by her enemies.

Or:

2.   She leaves because she and her husband have agreed that there are tax advantages in her taking a career break at this time.

Our evolutionary design is one that encourages us to prefer 1 to 2—to think in terms of human dramas, especially those involving power, deception, and negative emotions. It is essential that we be well informed about the greatest dangers in our world: other people. We need to understand what motivates them and how we should act to avoid danger at their hands. The stories we tell our children are just like this. They instruct about moral worth, treachery, friendship, nobility; they teach about the social landscape and how to navigate it. And as with all maps they are but simplified representations of reality. We know our children are going to have to play the confidence game we discussed in the last chapter. It does not help to believe that the world is full of happenstance and serendipity. That belief is demotivating. Heroic sagas help us psych ourselves up when we have to take on difficult tasks with courage and conviction.

THE EMPATHY DILEMMA

POTENCY IS added to our storytelling when we ourselves are part of the action. The whole point about narratives is to give us maps that help us navigate life. To practice our navigation skills, we need to be active communicators. Much of our face-to-face talk is idle chatter. It serves a number of vital functions, one of which is helping us gain and maintain facility in understanding and dealing with other people. This is everyday mind reading—intuiting what are the motives and feelings of people we deal with. At the same time, we try to maintain a tight control over how we expose our own thoughts. We don't want to leave ourselves vulnerable, but we do want to appear attractive and appealing to others. It's quite an art, and some people become very good at it. But mistakes also happen.

Jill interviews Jack for a job. They hit it off right away. She likes the straight way he looks at and talks to her. She feels she can trust him. It's not just that he seems smart but that he behaves in ways that feel right to her. She feels they are in tune with each other. When he smiles at something she says, she feels appreciative. His sensibilities seem quite parallel to her own, which is reassuring. When he nods his head in agreement with something she says, it seems to be genuine, not just a good interviewee act—he really seems to be listening with interest. When she asks him a question, the way he pauses to think makes her feel he is really considering her words. These reflections are only partly conscious; she would have difficulty putting them into words beyond a general impression that the chemistry is right. There have been plenty of previous occasions where she has known that the chemistry was wrong right from the start of an interview. Why, remember the guy she interviewed yesterday, Joe. What a contrast! He seemed gauche, phony, unspontaneous—someone who would irritate the hell out of her in no time flat.

Jill is probably right. She will work much better with Jack than Joe.

Jack gets the job.

This is what happens the first time they are out on a trip together to see a client.

Everything is going fine until the client makes a rather foolish remark and Jack seems to laugh at him. Why is he amused? The client has said something stupid, and Jack laughs! As they leave the building, Jill questions him. What was so funny? Jack mutters something to the effect that he wasn't laughing at the client, just being friendly. He casts her an odd glance, looking uncomfortable and rather ashamed. She, too, is embarrassed by the incident and decides to say no more about it.

Let's replay the scene from Jack's perspective.

As soon as Jack sees Jill, he knows the interview will go well. The way she looks at him and seems interested, even before he's said anything, makes him feel encouraged. He feels he doesn't have to put on an act—well, not the crude "buy-me-I'm-wonderful" stuff some heavy-duty interviews require. This is a situation in which he can sincerely be himself or at least act in an "I'm-being-myself" kind of way. He feels he can relax into the interview and just let it flow. She has an attractive smile, and he feels instinctively that he likes her and she likes him. All seems plain sailing. He knows he'll get the job.

At their first business assignment together, he feels positive and energized. The client seems to be a bit of a bore and not really on the ball. Jack figures he should just play along and try to stay involved in a friendly way. After the meeting, while he and Jill are on the curb looking for a cab, she suddenly turns on him and with a very unfriendly look in her eye asks him why at one point in the meeting he laughed at the client. He is astounded. He has no memory of doing any such thing. Besides, he would never act so unprofessionally in a meeting. So what's her problem? Maybe she's under stress or anxious about not landing the client. He wonders if he's really got her figured as well as he thought he had. He's going to have to watch out in the future.

Jack and Jill are victims of the empathy dilemma. It has two horns. The first is how brilliant we are at intuiting when we are in or out of tune with other people. Look at how skilled practitioners navigate their way around a cocktail party, easily slipping away from people with whom they have little in common and remaining longer with those who engage them. I have a friend who, after the breakup of his marriage, enrolled with a high-class dating agency. He found that he knew within the first ten minutes of each date why any relationship would have no future.

How do we do this? We are superbly equipped to read one another for signs of what we are thinking and feeling and from this to come to swift judgments about one another's trustworthiness. We do this by paying attention to two sets of indicators. First we try to determine the other people's emotional state through their tone of voice and general expression. From these we figure out whether they are warm, relaxed, angry, and so on. Next we look for signs of interest in us by the look in their eyes and general posture. We especially pick up facial expressions—down to the dilation of the pupils—that let us know whether or not they are paying attention, interested, alert, and friendly.

The second horn of the dilemma is that we often get this completely wrong. Interviewers use the same skills when they intuit whether a job candidate is someone they could warm to and work with. Unfortunately, when it comes to interviews we know this takes place much too prematurely, especially when the interviewee is working hard to convey an impression. Three minutes is said to be how long it takes most interviewers to form an opinion—hardly long enough to discover someone's hidden depths but time enough to experience an emotional reaction.

Errors in empathy occur for several important reasons. One is that the signals sent may be false. The codes for emotions in tone and expression are biological universals, and for this reason we can figure them out and fake them. This is a tricky art requiring a lot of control. Some people are much better at it than others. Sometimes we can spot the difference between good and bad actors, but it is not easy. One problem is that some signs are heavily influenced by cultural norms. Staring into an interviewer's eyes would be a mistake when going for a job in China, where it would be seen as impertinent and immodest, while in the United States averting your gaze would be seen as shifty and evasive.

The empathy dilemma comes from our strong desire to create harmony in face-to-face interactions. This means that unless we have some reason to be suspicious that we are being conned, we will incline toward empathy. It follows the logic of tit for tat that we encountered in the last chapter. So strong are our cooperative motives that we even engineer empathy through our body language. People unconsciously mirror each other's nonverbal gestures when they're getting along—putting their hands behind their head at the same time, leaning forward simultaneously. Watch young lovers facing each other over a café table to see this silent synchronized dance in a highly developed form.

It is also easy for empathic engagements to trap us. Perhaps you've had the experience. You strike up an interesting conversation with a stranger at a party, and just when you're warming to the interaction with him, he says something that completely deflates your view of him. On a grander scale, it happens with the sudden falling into and out of love of teenagers. Just when Brad and Brenda have gotten to an advanced stage of telling each other how perfectly aligned they are in their tastes and interests (love talk involves lots of this sort of empathic conspiracy), they go to a movie together and the following happens: Brenda is walking out of the cinema with a lingering glow of reverie from the wonderful film, and Brad turns to her and says, "God! That was truly awful, wasn't it?" Some relationships survive even this. The empathy instinct will paper over the cracks and steer people on to a safer territory of shared interests.

This is Jack and Jill's problem. They are victims of the empathy dilemma created by our speed-reading of each other and our desire to do deals.

SPEAKING AND LISTENING: HOW MESSAGES MISS THE MARK

HAVE YOU ever thought you had agreed on something with someone and the deal unexpectedly fell apart? Each of you thought the other was going to do *x*, so you both did *y* and *x* never got done. A simple miscommunication. Life is full of such instances, and in organizations they are routine.

The empathy engine works fast on strangers, but once people have passed our trust test we quickly economize on effort with them—we take their reactions for granted. We are equipped to make rapid judgments about other people that, if favorable, create empathy, understanding, trust, and goodwill. This is generally useful because it enables us to perform confidently. People who communicate with confidence get better deals and receive more positive regard. The cost is that trust decreases our alertness to negative signals; indeed, we tune them out when we sense that they may be harmful to our performance. People who are perpetually suspicious of others make poor communicators.

Errors do occur. If we are free from having to worry about whom we are communicating with, we will be giving our main attention to what we want to achieve through what we say. Each of us is, quite naturally, locked into our own worldview and its primary concerns, not what others around us want or expect to hear.

Most of the time in face-to-face communications, self-centeredness is regulated by the immediate feedback we get from each other. The process is incredibly fast, unconscious, and highly skilled. We instantly detect through facial and verbal expressions when a person we are talking to is missing the point: the face sends a signal. It may only take the slightest of frowns to indicate that you don't understand or agree with what someone is saying to you, and they will change tack or ask you what's the problem.

There are big individual differences in this sensitivity. Some people are especially obtuse and seem not to notice or even care what others think and feel. This is partly a hardwired personality factor. It also varies as a function of power and status. Dominant individuals often seem deliberately unconcerned with others' reactions to them. Conversely, people in highly vulnerable and dependent positions tend to be supervigilant in their reading of signals from others, often to the point of anxiously overinterpreting ambiguous signs. It is, incidentally, the desire to convey the appearance of being unconcerned with others—a mark of status and power—that motivates the affectation of looking cool by wearing dark glasses in gloomy weather and nightclubs.

E-BABEL: LIVING WITH THE NEW COMMUNICATIONS MEDIA

TAKE AWAY the visual feedback in communications, and see what happens. Look at our behavior on the telephone, for example. The phone removes access to a lot of the usual cues that provide feedback and protect against errors. The absence of these factors shortens much phone talk to skeletal efficiency. To make most phone conversations last as long as the same talk would take face-to-face, you have to do a lot of the verbal equivalent of nodding and smiling—by saying "Uh-huh" or "Ah" at frequent intervals. Even then it is hard work for most people (except some teenagers!) to sustain a phone conversation for as long as you can talk face-to-face.

Thankfully, tone of voice and the immediacy of response on the phone usually prevent serious miscommunications from occurring. Moving into the realm of electronic mail takes us one step further away; even when we are exchanging with people who just sit by their machines waiting for our messages, we have lost the immediacy of instant response. Communication is serial. I speak for a while, sign off when I am through, and then wait for your reply. You can't interrupt me or interrogate while I am speaking. To all intents and purposes we are like old-fashioned letter writers. But with e-mail things can get rough very quickly. Most users will agree that it is frighteningly easy to get into conflicts and misunderstandings with people. The correspondents of the nineteenth century, knowing that by Pony Express it would be still some weeks before there was any chance of reply, took great care over the words they used, and wrote with elaborate etiquette. Even their penmanship signified the importance of the communication. Nowadays we don't even have to worry about the legibility of our handwriting; we can just fire off messages. In most networked offices you find people sending e-mails to others who are just a short walk down the corridor. Sometimes this is deliberate: when people are under intense pressure to process information and do their jobs, it is efficient—they know they won't get stuck in your office having a conversation.

Face-to-face communication has an automatic code of manners; the ways we sign on and sign off are analogous to the sniffs, grunts, and posturing of the greetings and partings of many other social mammals. Such niceties are lost in the e-mail world. Now we increasingly encounter a new phenomenon, what we might call "e-rage," the electronic equivalent of road rage. The problem starts because there is no warm-up to frame your intentions. You send off what you think is a simple message and get by return an unexpectedly sharp response. You fire off an equally sharp response (tit for tat), and before you know it you are in a minor spat with a colleague. People get very upset. Once the face-to-face cues of people's intentions and emotions are removed, and as there is no tone of voice to reassure you, there is much more room for misinterpretation. The message sent is not the one received.

The preconditions for e-rage are similar to those for road rage:

◇ *Depersonalization.* The other person is reduced to the message. It is easy for us to tell a story to ourselves about their intentions and motives as a simplified moral saga where we are the good guy and the other person is just another vehicle behaving badly on the information highway.

◇ *Implicit rules.* Norms and expectations govern conduct; an implicit contract of expected etiquette hangs in the air. Just as we get angry when someone else drives rudely, we get annoyed at messages that lack niceties.

◇ *Ambiguity.* We have no way of checking the meaning of ambiguous signals or behaviors.

◇ *Self-centering.* In the absence of feedback, we find it difficult to decenter, to see how things might look different from another person's perspective. We forgive ourselves when *we* break the rules, however, because we can see perfectly well the extenuating circumstances that justify our own behavior.

The generally hasty, ill-considered, and uncrafted nature of electronic communications makes these kinds of problems all the more likely to occur. Even professional writers have to work hard to read their own words as if they were readers. One trick for doing this is to print out and read your own messages before you send them, as if you were the receiver, not the sender. This device for decentering isn't a guarantee against problems arising, but you'll be amazed at how often you'll spot something that reads worse when you see it as hard copy in your hand than viewed as words on your screen.

Many avoidable problems come from this fundamental principle of communications:

It is the receiver, not the sender, who controls the effect of communication.

It doesn't matter how skilled you are as a communicator; if I don't trust you, even your most sincere words will sound suspicious to me. If I am bored and tired, you will have to pull out something special from your communications armory to get me to sit up and pay attention. If you send me a message that can be interpreted in more than one way, I will interpret it in the way that makes most sense from my point of view—not yours.

Reading your own communications as if you were the receiver doesn't guarantee success, but it may save you some hassle.

GOSSIP AS GROOMING

IN NONHUMAN primate groups, individuals spend a lot of time paying elaborate physical attention to each other, doing such useful work as picking fleas and lice out of each other's fur. This is not, however, a random or democratically ordered system. Grooming is practiced asymmetrically; it cements bonds. Parents do it to offspring, but not vice versa. Low-status members do it to high-status members. Individuals who want to curry favor or repair a quarrel do it. Grooming is part of the politics of the troupe. Likewise, the food sharing of chimps is not arbitrary, it is political. One with a surplus to share will be selectively generous, giving to build coalitions, enhance reputation, repay favors, and escape persecution.

In human societies we share information as chimps share food. A person standing in the middle of a room regaling colleagues with a juicy story is sharing a vital human resource: gossip. A person huddled in a corner with a colleague, telling him some piece of minor sensation about a common acquaintance, is grooming—letting him know he is important and liked enough to be trusted with a confidence.

It is a common misconception that gossip is a female pastime. Anyone who believes this has never been to a conference, trade fair, business luncheon, golf club, or any of a thousand other places where men congregate with their peers. It's just that men can get away with it by calling it "networking"—a term invented to make male gossip respectable.

Look at what happens at a professional conference or large business meeting: If you walk around eavesdropping on what people are talking about in their various twos and threes, what will you hear? Are they earnestly solving work-related problems, getting technical information, taking notes, and learning new facts? When they get home, that might be what they tell their wife, husband, or boss to justify the time and expense of their absence. But you just have to walk around a little with your eyes open to see what is really going on. People are having fun, for goodness' sake! You can see it in the animated way they talk to one another. And what are they having fun doing? They are telling stories—about themselves and mutual acquaintances. They are gossiping about the famous and the infamous. They are reprising stories of times they have shared at previous meetings. There is a good deal of smiles, laughs, and excited exclamations. Heads are bent close together. Eye contact is intense, denoting absorbed interest. Little pieces of theater are enacted as one person holds forth in entertaining style to a surrounding group. And this is just during work time. In the evenings the local restaurants and bars are crowded with delegates continuing furiously to oil the network with the aid of large quantities of food and drink.

Even when people are talking about operational matters,

genuinely learning and finding intellectual stimulation, this does not prevent there being a more primitive social agenda. For example, Frank is quizzing Pedro about a project Pedro's firm has been engaged on. What Frank is really interested in finding out is:

◇ Was it a success or a failure?
◇ Is Pedro now a more or less valuable person to know as a result of this?
◇ Is Pedro's firm doing better than Frank's firm?
◇ How does Frank stack up against Pedro on his career achievement timetable?
◇ Should he take at face value what Pedro says about the current business situation?
◇ What is going on at Pedro's firm that might affect Frank at his?
◇ What is Frank's external reputation? Are people at Pedro's firm aware of Frank and his work?
◇ Who are the important newcomers in their field, and whose stars are fading?

Quite apart from any valuable factual information it might provide, such chitchat fulfills a number of other vital functions:

◇ *Status checking:* They are establishing where they stand relative to each other in the constantly changing hierarchy of their professional community.
◇ *Cast-listing:* They are compiling and checking information about who are the key people in the network, from newcomers to old-timers.
◇ *Weather forecasting:* They are exchanging data about the location of potential threats and opportunities that might arise for either of them.
◇ *Line laying:* They are resurfacing an information highway between each other, so that next time they

meet at a conference they will be able to pick up where
they left off and get the latest news.
◇ *Grooming:* They are telling each other that each is
someone who is worth talking to and that they like each
other's company, thus affirming each other's value as
people of high standing and good character.

This is a typically male agenda. Research has shown that
men and women gossip equally, but in different ways and with
different emphasis. Men are obsessed mainly with mapping status
so that they will know where the action is and can figure out
whether they should play the game differently. Women are mainly
building relationships, establishing connections that will keep
them involved, supported, and emotionally engaged. This division
of labor is not exclusive, however; both sexes need to fulfill both
these functions.

Indeed, scientists have come to conclude that this is actually
the main evolutionary purpose of our braininess.

THE RULE OF 150 AND THE
MACHIAVELLIAN INTELLIGENCE
HYPOTHESIS

EVOLUTIONARY BIOLOGISTS have concluded that the enlarged
brain of social mammals—the monkeys and other primates—
grew to its bloated proportions precisely for this purpose: to facilitate complex group living. It surpasses the mental hardware of all
other species to enable us to get along in larger and more complex communities.

As we have seen, the brain is not designed for sitting around
doing mental arithmetic. Even our wonderful visual system and
unique speech centers do not account for the size of the brain.

The clue is to be found in the main structures that set us apart from all other species: the bulbous lumps at the forward end of the skull called the frontal lobe area. The mysteries of these organs are still being unraveled, but this much we know for sure: the enlarged cortex gives us the capacity for a high degree of conscious control over our emotions and their expression and is indispensable to the functioning of our elaborate memory systems.

These features are consistent with the idea that the brain grew over the succession of hominid species in an "evolutionary arms race" to be better than one's neighbor at what was most key to survival and passing on one's genes: being smart socially. This means: a better mental map of the social network, more intelligence about whom to trust, a more complex set of deals, more persuasive powers for forming relationships, and better social navigational skills. All these enhance one's chances of acquiring a good-quality mate, ample resources for oneself and one's family, and the best start in life for one's offspring.

There is a limit to these capacities, though. For our ancestors the primary community would have been a seminomadic kinship group, an aggregation of loosely interrelated extended families. If a community like this becomes too small, it becomes vulnerable to resource depletion, too large, and it cannot support itself. In between is the group size we can comprehend and adjust to. Our brains need sufficient capacity to help us do this. Indeed, this would have been a significant driver for evolution of the brain: creating an organ big enough to manage in a group that is of the optimum size for survival. This idea led psychologist Robin Dunbar to try to determine whether, among various primates, there is a relationship between the size of the neocortex and the average size of communities. Sure enough, there is a remarkably strong and clear correlation, from the smallish packs of the not-so-bright monkeys to the large troupes of very smart baboons. The next logical question was, if the dimensions of the human brain are entered into the equation, what indication does it give of the "natural" size of the human community? The answer is about 150, the scale of the social network that our brains can contain and navigate.

How many people could you borrow money—say $20—from without specifying when you would pay it back? How many people in your address book do you keep tabs on from year to year? How many people at your workplace or in your professional network do you have conversations with on repeated occasions? What's the size of your active e-mail network? Add these numbers together and the total won't be exactly 150, but it is unlikely to be much larger. Dunbar looked at a variety of historical and anthropological sources and found that this number crops up time and again for the size of human subclans, rural communities, fighting groups, personal networks, and the like. Dunbar notes that the number of people whose faces we can put names to is much larger—up to 2,000, which, he says, is "suspiciously close to the size of many tribal groups." Tribal groups, he notes, are typically aggregations of many such 150-member clans.

When we gossip, these are the proportions our networks reduce the world to. We might work in much larger units, but the important psychological community with which we engage is much more local, even if its members are scattered around the world. This is the size of the theater in which we perform socially. Good performance, as we have noted in previous chapters, involves being able to play competitive games of social advantage and cooperative games of alliances and bonding. It involves knowing how to create a good impression, spot cheats, and keep tabs on key individuals. It also requires perfecting the skills that enable people with "bad" memories to beat at chess computers with "good" memories: the arts of misleading, setting traps, making strategic sacrifices, and using other get-ahead tactics. We have also seen that when we are dealing with real people we need to practice these arts with a clear conscience and without loss of our reputation. Having to turn in an unblemished performance while on stage with other humans playing the same games makes it necessary for us, from time to time, to deceive ourselves about our own true motives and abilities.

This bundle of dramatic arts and mental skills is what evolutionary scientists call the Machiavellian intelligence hypothesis. It is invoked to explain not just humans' but all primates' mental gifts, for we are not alone in these practices. Political games are being played in the monkey or ape colony in your local zoo at this very moment.

THE POLITICS OF THE NETWORK

L ET'S GO to a human zoo, the conference, and see what is happening. Even at a huge professional association meeting with thousands of delegates, the Rule of 150 will be operating. Some attendees may know many more by name or sight, but even the most addicted conference will not be chummy with more than this number. Since every individual's network is interconnected with others', it is important to have the right people on one's own network. That way you can connect, at least indirectly, with the key players. The whole purpose and design of the big conference is based on this understanding. Participants are given name badges so that as they walk around they can continually scan the network. Conference presentations are scheduled so as to maximize opportunities for individuals to display their expertise to one another. Various events, from keynote presentations to small business meetings, are designed to allow people a variety of ways to pick up news and enlarge their knowledge of who stands where in their profession's hierarchy. Award ceremonies and formal speeches mark the peaks of the status landscape. Parties, bars, restaurants, and foyers are thronged with gossiping small groups and grooming couples.

A walk around will reveal how well embedded people are in the network.

First are the *beginners,* people not long on the conference circuit. They know that no one knows or is interested in them. They are here to put faces to the famous names. They walk around with their eyes focused at chest level, scanning badges. When they see a name they recognize, they look up at the face and press a subliminal mental "record" button. They dutifully attend all the sessions and diligently take notes. There is nothing else to do. Mostly they will be ignored, unless they are attractive females. A shy person can have a very lonely time at these events. Bolder beginners will try to strike up conversations with strangers. This will meet with some success, especially with other beginners, who, as people with nothing to lose, cannot afford to be rivals; they have more to gain by trading insights and the comfort of friendship. If beginners approach the veterans and the famous, they will often find them to be friendly and magnanimous. This is costless for the famous, for whom it even has a benefit: it is good for their self-image and reputation to patronize the lowly, especially if there is any kind of audience of more important people to bear witness.

The *insiders* are the group with whom the beginner can expect least contact. They are people in the middle ranks of the community, who are competing for recognition. They scan badges and network with an often furious intensity. They spend only a moderate amount of time attending the formal presentations. The real action for them is where they can meet and talk to people they already know and get acquainted with high-status people they don't; the cafeterias, bars, bookstalls, and lobby areas are thronged with them. These insiders are the main drivers of the network. They have come to the conference to publicize themselves, expand their network, and get recognition. The extroverts among them will work hardest. More laid-back and introverted members of this group will prefer hanging out in the comfort zone of a familiar and friendly gang for the duration of the congress.

Finally, there are the *famous* and *nearly famous*. Some of these can't stop themselves from acting like insiders—as if they were still on the way up or might slide back down the pecking order. It a common compulsion of people who have reached the highest possible position of achievement or status to carry on as if there were still further stages to go. The truly famous know that they simply have to act in role. They can afford to "forget" to wear their own badge and ignore others'. They can simply focus on the faces of people they know and like, and the rest of the world will come to them. Their contributions are in events either where they are the center of attention or where they are honored guests, appearing as a favor to a friend or protégé. They do not stay for the entire conference. It is a mark of their status that the conference needs them more than they need it.

Network politics are being enacted everywhere. Here are some of them.

Power Displays

These are blatant. For example, Charles is standing in the middle of the bar. His voice is booming—louder than it needs to be to reach the ears of the four or five people standing around him. He is recounting an episode in which he was the comic hero who committed an amusing act of folly. This story is mildly self-mocking; he is playing the game of showing his strength by being big enough to admit a weakness. His audience has varied motives; they are a mix of allies, aspirants, and dependents. The *allies* are working the crowd with him; they are his straight men (or women). The partnership is mutually beneficial. They get some status by association, and Charles gets to bask in their good regard. The *aspirants* want to see how it's done. They watch with fascination, trying to pick up tips and to spot signs of weakness. Anything that reduces the gap between them and Charles will be welcome. Sometimes they may lob in some challenging remark to see how he reacts. The *dependents* are like the allies, except that they have no status or power and little immediate hope of getting either. Like the aspirants, they are pleased if the distance between them and Charles is reduced. So if Charles gets drunk and begins to make a fool of himself, the aspirants will drift away but the dependents will hang in there, getting the double benefit of acting out a buddy role and seeing him reduced in stature.

Assassination

This is not always easy to detect. Eric tells Carol the inside dope he has on Matt. Eric knows this is a risky ploy, as malicious gossip often rebounds on the teller. Isn't Carol likely to reason that if Eric can say this sort of stuff about Matt he might say the same kinds of things about her behind her back? Crude assassination attempts are likely to be made only when the teller believes that his listener also would like to conspire in the victim's downfall. Thus much blatant malicious gossip is against a common enemy, such as a boss, outsider, or common rival. Thus if Eric wants to strike against Matt, his attack will be well disguised and stealthy so as not to appear malicious. To this end, the story Eric tells Carol about Matt sounds innocent, even affectionate, but has enough of a spin on it for Carol to be privately shocked and for Matt's reputation to have been dented. A few more hits, and he'll sink.

Rutting and Strutting

This is easy to spot without hearing any of the words; the body language speaks for itself. Dennis is standing far closer to Rachel than he needs to. His eye contact is prolonged with lots of smiling. Occasionally he touches her lightly on the arm—just to emphasize a point, of course. She is smiling and nodding a lot, encouraging him. He is doing most of the talking. He is selling something: himself. In another corner a similar game is being acted out. A little group of young men are surrounding Linda. She is talking excitedly with lots of hand gestures and head tossing, inclining her body this way and that in the direction of the men around her. Welcome to the sexual marketplace. Some are playing the game for real. Numerous marriages owe their beginnings and ends to just such conference liaisons. Many are just play-acting, trying to determine whether they are still attractive. This does not have to be at a physical level; in many contexts the flirtation is intellectual: you tell me you like my ideas, and I'll tell you I like yours.

Skirmishing

This is going on everywhere. Sometimes it's obvious, sometimes less so. For example, Sam and Leo are friends, as is obvious from the kind of shorthand speech they use and the familiarity of their jokey insults of each other. These boys are wrestling, a serious exercise before an audience. They affect to be interested only in each other, but they are grateful to each other for the opportunity to show off their mental skills and resilience and glad to have an audience of bystanders. Without an audience the sport would lose much of its pleasure. Elsewhere in the room other skirmishes are going on, mainly among the men. In some cases the fighting is bruising. Ted, Paulo, and Alex—none of whom knows the others well—are arguing about a technical point from a presentation all three just attended. Here the fighting is serious. Each cares a lot about winning his point. No one is laughing, except to disguise hurt when they've taken a hit. They are talking fast, constantly interrupting one another. The flow of conversation is complex. Ted briefly seems to agree with Paulo, but the next minute he switches to trying to prove him wrong about something. Alex struggles to make his points and eventually drops out, leaving the main fight to the other two.

Bonding

Finally, some groups just seem to be having fun. They are bonding by trading. Each person gets a turn at telling stories and being listened to by the others. As the conversation rotates, everyone seems to be striving to maintain a friendly, attentive, and positive demeanor. This is a manifestation of two of the key gratifications of being in groups: belonging and display. People get the chance to try out self-presentations in a context of acceptance. They get the deeply reassuring feeling of liking and being liked by relative strangers. It is a network reality check. Karen has just joined the group. As she listens to the others, she gets a sense of the reality of these other people's lives and how it compares with what is going on in her own life. When it's her turn to speak, she finds she likes the feeling that people are interested in her.

These micropolitics can be seen in all large gatherings of people who have a shared interest: trade fairs, major company meetings, conventions. It is no exaggeration to say that such events exist only for this purpose. Networking is important work in its own right and as old as the human community as a means of regulating the culture.

On a wider stage, the human obsession with status ordering dominates business journals and other media in the form of rankings. In the marketing profession there are numerous awards so that agencies can continuously compare themselves with each other. Business schools obsessively read the media ratings and compete with one another to improve their standing. Blue-chip companies vie for recognition from the business press, finance analysts, and consumer bodies. This is what you get when you put together the human predilection for categorical comparisons and hierarchical ordering in an era of mass media.

RUMOR IS IMPROVISED NEWS

OUR EVOLUTIONARY heritage has made us acutely attuned to uncertainty and danger—in fact, to any kind of threatened loss. Our bodily systems respond immediately for fight or flight in the face of visible danger. But what if the threat or uncertainty is not obvious, but just a nagging feeling, a growing sense of unease because of something one has seen or heard? You search for news.

Franco doesn't realize it, but he is about to set a rumor running. He is doing some photocopying when he hears two managers talking conspiratorially in the corridor outside, unaware of his presence. One of them, James, is saying to the other, "It's going to be a hard winter—I don't know where we're going to find cost savings." Franco is alarmed. As a recent hire he is well aware that the firm operates a LIFO—last in, first out—layoff policy. Faced with this kind of uncertainty, Franco cannot figure out anything specific he might do to help himself. But he can act on his feeling, and that feeling is anxiety. The way to deal with this is to get more details. So during the lunch break he goes to Lenny, a friendly old guy who has been with the firm for years. Franco asks him, "Have you heard anything about the company planning to make cuts this winter?" Lenny hasn't heard this, but says he's not surprised. As an old-timer he has witnessed such events before. His narrative motive kicks in; the story makes sense of other things he's noticed. A long-term customer cut its order this year. He's been reading in the papers of a possible downturn in the industry. It all adds up. Lenny feels that this information gives him a certain responsibility. When he meets his line team the next morning, he gravely announces, "You'd better rethink your vacation plans and cut down your Christmas list. It's a pretty sure thing that we'll be on short time, or there might even be layoffs. Quite frankly, I wouldn't be surprised if they close down or sell off our bit of the business." It is not long before this story gets back to the managers Franco overheard in the first place. James hears from someone that word is going around that the company is in financial difficulties and planning redundancies. James assumes that this is independent confirmation of his own fears. It never occurs to him, of course, that it was his own overheard conversation that set the rumor running in the first place and that it has since traveled a long complete circuit of Chinese whispers. Like many others in the company, James now concludes that top management has a secret plan that will put jobs on the line. The first thing an astonished top management hears about all this is from the HR manager (one of the few executives who do wander about the plant talking to people). At the next executive board meeting he tells his bewildered colleagues that there is a widespread belief that the company is in big trouble and is planning to close at least one of its plants.

Nature abhors a vacuum, it is said, and communities abhor an information vacuum when there is uncertainty or threat. Because of our need to have a sense of control, we clutch at straws of information and create explanations that make sense of them. Once we see one piece of a jigsaw, it is natural to sketch in your mind the shapes of the surrounding pieces. You could be right. You could also be completely wrong.

Of course, that's the problem with rumors. Rumors often prove to be false, but they do not need to be confirmed more than once for people to make a psychological bet on their being true. In other words, you might as well assume the worst until you hear the all clear. The alternative option—assuming a rumor is false and then getting caught out—could be suicidal. This is a further extension of the loss aversion principle. We were designed for survival in a dangerous and uncertain world. Our ancestral environment was quite different from our own. Death could strike

unexpectedly from many sources. Food supply could be inter-
rupted at any time. Daily acts of survival were full of hazard.
Believing news, in this context, is prudent. It is vitally important
to know what danger one could be in at any time. The line into
the gossip factory keeps one informed about how close threats
might be at any time. And even if nine times out of ten a rumor
proves to be false, it would be foolish to ignore the signals and
bet one's life on a 10 percent chance. The fail-safe option is to
pay attention to them all and make provision for them as if they
were true.

In today's world, so much is within our control that disasters
rarely affect us directly, but our psychology still remains tuned to
a world in which they might. And actually, our gossip instinct is
continually being reinforced by the misinformation and secrecy of
leaders and officials. If your plant is about to close or a new highway is going to be routed through your backyard, then you are
most likely to hear about it informally from someone you know
before it comes through an official channel. No one likes to be
alone with fear, and spreading bad news is one way of creating
the comfort of sharing. There is some kudos to be had by being
the first to tell other people important things, especially bad news.
It shows we have status as people close to sources of knowledge
and centers of power.

We are obsessively interested in and attracted to bad news. If
there is a fire in your neighborhood or a major road accident on
a highway, people are drawn to watch in curious fascination.
Even if our rational minds believe such calamities won't happen
to us, we still want to know what they're like. We may even catch
ourselves taking secret pleasure in knowing that a calamity has
happened to someone else and not to us. The Germans have
a wonderful word for this: *Schadenfreude,* or joy in another's
misfortune.

Today we also face a new problem: we are drenched in

news, most of it about things that will not touch us directly. How are we to sift out what matters? Keeping up with the sheer complexity of it all is a struggle. Some people obsessively soak up all the news and current affairs they can get from the papers and TV. Some of this is prudent—like watching the movements of stocks to protect one's investments. But most of it has a more general inquisitiveness motivation. Newspapers are full of "human interest" stories and reportage about the famous and infamous. This is glorified gossip that feeds our network needs and helps us keep tabs on social trends—to gauge what is "normal" and where we stand in relation to the rest of our clan. It also supplies useful material for grooming—something to talk about with others. Armed with what we glean, we can show ourselves to be well informed and hold our own in conversation as equals with other well-informed people.

Managers are no less interested in bad news than their staff, but when they think they might be blamed or thought badly of, their instincts quickly move in the opposite direction. On the loss aversion principle, they would prefer to shut their eyes and pretend a problem doesn't exist. Denial, the ultimate avoidance technique, is a mixture of magical wish fulfillment and the confidence illusion. Locked into their own perspective, executives want to believe that any crisis can be downplayed and that things will get back to normal before anyone is blamed. At the corporate level this kind of avoidance is disastrous.

Consider the contrast between TWA's and Swissair's reactions to two separate but similar plane crashes. In both cases all lives were lost when passenger jets bound for Europe crashed into the sea off the U.S. coast.

The TWA loss in 1996 coincided with the recent departure of two top executives and the company's CEO being away in Europe on business. This created a gap in top management's decision making. Additionally, security concerns about possible terrorist implications obstructed the free flow of information. The way the company handled the crucial first twenty-four hours after the crash aroused great anger among friends and relatives of the deceased. The company refused to produce a passenger list until all the families had been notified. It didn't sufficiently staff the telephone hot line it established for relatives, leaving many of their calls unanswered. It paid little attention to the travel and accommodation needs of the families. There were accusations of incompetence and insensitivity, including trenchant criticism from New York City's mayor, Rudolph Giuliani.

Two years later, Swissair (with the aid of its partner, Delta) responded to a similar disaster with speed, efficiency, compassion, and generosity. Within hours a full passenger list had been published. Families were immediately offered $20,000 to cover expenses. Fully functional hot lines were set up. An army of crisis counselors was drafted in to work with grieving relatives. These prompt actions earned the company much praise and enhanced its reputation.

When disasters occur, executives' immediate response has major impact. Their actions in the first twenty-four hours afterward are crucial. As one commentator noted in dealing with such PR disasters, "The earlier you write the checks, the smaller those checks are going to be."

Increasing numbers of companies are now espousing openness. The word figures in many businesses' mission statements. But the test comes when really bad news occurs. Fear of loss of control is likely to take over, rendering top management too terrified to walk their talk of openness. A little decentering would tell them what their employees and others need to know in a crisis, and a little foresight would tell them what will happen if they don't allay their anxieties with information. People will make up the news for themselves and mistrust their bosses forevermore.

WHY AND HOW ORGANIZATIONS FAIL
TO COMMUNICATE EFFECTIVELY

COMMUNICATION IS the lifeblood of an organization. If you study how it flows internally you can diagnose all the strengths and ills of a business. You will discover: Does the company have a clear strategy? Do managers have confidence in the strategy? Do people trust one another? Does influence flow easily or with difficulty? Are leaders respected? Are groups in conflict with one another? Are decisions being made rationally and effectively? Is the culture cohesive?

Every year I take parties of executive MBA students on overseas field trips, where they do free miniconsulting projects for various companies. Teams of six students conduct "communications audits" in firms of all kinds—large and small, public and private. Dividing the labor among its members, each team does three days of intensive interviews and conducts focus groups at all levels of the business, asking about internal communications: formal and informal; up, down, and laterally. On the fourth day team members retire to their hotel to prepare a presentation for delivery to the company on the fifth day. It is amazing how quickly and accurately they capture the essence of the issues facing the business within just a week. This happens not just because the students are intelligent but because most of the time no one goes around a company asking the following kinds of questions and listening to the answers:

◇ About what kinds of issues and areas would you like the company to communicate with you?
◇ By what media would you like your managers to communicate with you?
◇ What messages from the company reach you, and what sense do they make to you?

◇ What would you like top management to know about your situation and how you feel about it?

Asking questions like these, in the role of outsiders gathering information in guaranteed confidence, is the equivalent of doing an MRI on the body of the business. And intracompany communications reveal the best and worst of a human organization. This kind of inquiry confirms the Tolstoy Principle (slightly modified) introduced in Chapter 1: that good communications are the same in every firm; bad communications differ in every firm that gets them wrong.

The companies that propagate the best communications are founded on trust and community at three levels:

1. Key positions are occupied by communications champions—people who revel in talking and listening to other people about things that matter to them.
2. When employees congregate in groups, they do so with confidence that they will share information and in a spirit of good-humored trust.
3. The company practices the core values of true community, spreading by example communications practices that embody the values of openness, involvement, and recognition.

Sadly, many more companies get their internal communication wrong than right, and not because they don't care. They just make rudimentary mistakes and suffer from a familiar range of ailments:

◇ Messages flow from the top and never reach the bottom.
◇ Rumors are trusted more than official communiqués.
◇ Employees do not hear from their managers about the things that really matter to them.

◇ People with information or critical issues fail to communicate clearly or on time.

◇ Employees feel that no one at the top hears or understands their concerns.

◇ People at lower levels have good ideas, but no one ever asks for them.

Consider the following case:

The chief executive officer of a major financial services business told me, with an air of resigned bewilderment, how he had learned from a recent employee attitude survey that he was perceived as being too remote and should spend more time in the branches talking to staff. Sound advice, but what happened was that whenever he dropped into a branch an air of panic ensued. On his unexpected calls, instead of warm and friendly exchanges there was stilted formality. His arrival created a great flurry as an office space was cleared for him. When he would sit with a local staff member in her office and ask how things were going, instead of frank insights he would get a lot of agitated reassurance mixed with polite small talk. The branch managers evidently dreaded his visits. Some staff were more appreciative of the fact that he cared enough to come by, but they saw his tours as basically a cosmetic exercise. "I felt like I was conducting royal visits," he said miserably. Shortly afterward, he gave the visits up.

The problem was that his behavior was abnormal for the bureaucratic business he headed and his attempts at informality just accentuated the sense of incongruity. You can't create an open-dialogue community by such simple means. The CEO's attempt was halfway right. In less uptight organizations, walking around is part of the culture and communications are conducted on more than one level. Memos and other formal media are augmented by managers walking the corridors and talking to people. It is accepted. Informal communications with top managers are not exceptional events; they are conversations between members of the same clan. It is quite a different matter when the chief executive of a ten-thousand-strong business drops in on unsuspecting people who have never been to the head office or seen him in the flesh before.

But contact between top management and employees can work, even in a large company, if a family culture is built from the roots of the business. In 1999 the South African retail grocery chain Pick 'n Pay employed approximately twenty-nine thousand people. It was the tradition of its chairman, Raymond Ackerman, to visit all 265 of its stores at Christmas time with his wife, shaking hands with as many staff as possible. These tours, though fleeting, were much looked forward to by employees. All management systems reinforced the idea that staff and customer relations were primary. A community spirit was emphasized at the store and locality levels by high-involvement decision-making. The company's personnel system underpinned its humanistic value set. The company was one of the country's most admired and successful businesses, built on and sustained by best-practice communications.

It is easier, perhaps, to achieve this in retail than in financial services. In banking many customers are more invisible and money is what matters. A manager with good accounting skills and poor communication skills stands a better chance of being promoted than one with the reverse combination.

The root cause of the fact that corporate internal communications are usually awful is that most businesses are institutions designed to control tasks and operations, not to function as communities. Understandably, this makes managers see communications as part of the control system. Messages are sent out to get things done and feedback is only practiced to check whether things have been done. In such an environment managers may be inclined to believe that networking is politically unethical; that rumor gives people false ideas; that gossip wastes time.

Many big businesses are placing increasing importance on transmitting "cultural" messages through their mission statements in order to help staff understand and share the company vision. These are laudable, but often they miss the point. Instead of standing for genuinely shared values in the company as a community, many of these mission statements are simply wish lists, attempts to put positive images into employees' heads. This often results in mission statements being perceived, with some justification, as little better than brainwashing attempts.

So what is the best thing to do? The answer is to create the structures and culture of a true community.

• 8 •

Life in Camp: Structure, Culture, and the Future of Organization

THE NATURE OF ORGANIZATION AND ITS CONSEQUENCES

M ANY LIVING creatures do well by congregating in large numbers. Thundering herds of bison traversed the wide-open spaces of America before humans hunted them to near extinction. More durable have been other huge groupings, such as termites, which divide their labor to build and maintain their turreted empires, or colonies of bees that communicate by aerial displays to direct the swarm to the next source of pollen.

Humans' efforts to work together sometimes look a little haphazard against these miracles of precise coordination. The reason for this is that we organize by acts of individual will and choice, guided by personal motives and perceptions, not by the inflexible preprogramming that guides less intelligent creatures. Colonies of ants and bees have a genetic identity that makes them the equivalent of superorganisms, with encoded instructions that tell each member what part it will play in the life of the community. Herd mammals are of higher intelligence, but basically they are still large family groups that stick together for the safety of feeding and breeding in numbers. Intelligent organization really starts only with the higher mammals of land and sea. Monkeys are truly social in a sense that cattle are not—because they form friendships and alliances, have rivalries, and organize themselves in different ways for each new challenge they face. This requires a great deal of brainpower, and, as we have seen, by the time the level of sophistication of humans is reached one has entered the realm of mind reading, deception, intuition, love, trust, jealousy, altruism, and psychological debt. These are the social arts that enable us to build alliances, identify enemies, choose mates, and work together.

The result is that when humans organize, an amazing variety of things can happen. We have developed social systems—conventions for mating, labor, and governance—of one kind to live by the fruits of the seas, of another to forage through mountainous terrain, and yet another to follow herds across frozen tundra. We organize in one way for a life of settled peace, another for warfare, and yet another for nomadic trekking. There might seem to be no limit to how we can organize. But in fact there are limits, and no way of organizing is without its problems.

We are communitarian creatures who regulate our relationships through culture. In this we differ from other species in degree, not kind. Chimpanzees have culture—different ways of living adapted to various habitats, including local traditions of tool use handed down from one generation to the next. Humans' adaptations to their environments are more varied and complex, but spring from the same two interconnected driving forces: the need to optimize mating opportunities and the need to secure essential resources. Anthropologists have found mating systems to vary for these reasons. Patterns of marriage, inheritance, dowry, governance, and sexual mores alter to deal with the availability of men and women to do the essential tasks of a community.

The way in which we organize is always a compromise

between our unchanging nature and the changing world. All organizations reflect the constant pressures of our needs and instincts. The variations among them accommodate local conditions and history. This makes every organizational form a kind of experiment—a working arrangement that mediates between the two pulls of human nature and human economy. Each version delivers some benefit and exacts some cost.

Hunter-gatherer existence might have had quite a lot going for it in terms of social integration, but the costs were heavy in terms of human vulnerability to a variety of hazards and ills. What came after this way of life—farming communities, city-states, and feudal nations—gave us new power and security from environmental threats, but they exacted a price, sometimes a very heavy one. Now humans had to contend with unprecedented conflict within their own ranks, including slavery and genocide. New health hazards resulting from unregulated mass living, gross economic inequalities, and malnutrition were visited upon the human race.

In the last two hundred years we have been enjoying a golden age. Through ingenious inventions allied with new forms of organization for work and governance, we have created a great bounty of resources and tools for living, as well as solving many long-standing problems that have dogged humanity for most of its post-hunter-gatherer history.

We have avoided some of the hazards and horrors of our forebears, but we have unleashed a range of new costs—many of them quite subtle and insidious compared with those of former times. Some of these were documented in Chapter 2 as the Seven Deadly Syndromes of organizational life.

As we enter the twenty-first century we face a vast magnification of benefits and costs. The information revolution promises new wonders for work and social intercourse. But the dangers are increasing, too. More and more people are suffering from modern plagues: feelings of alienation, loss of meaning in life, depressive illness, harmful addictions, and social conflicts. New difficulties are arising both between people and within individuals' minds. What lies ahead?

THE EP CHALLENGE

A SPECIAL FEATURE of the intelligence of the human animal is that not only can he talk with other humans, he can also converse with himself. This means reflection, planning, and adjustment. This gift is what has created the rich diversity of culture down the ages. It also has created many of the problems we face, by leading us through the following process:

1. We are quick to spot new opportunities for satisfying our impulses.
2. We are ingenious in thinking up new ways of organizing and working to exploit opportunities.
3. If the prize is big enough, we persist with these behaviors, even if they create new conflicts with other aspects of human nature.
4. We develop strategies for tolerating misfits with our surroundings and making do when we are in uncongenial circumstances.
5. We are often unaware of the delayed and long-term costs of our strategies for tolerating misfit (especially mental and physical health hazards).

In other words, our adaptive genius—our almost limitless capacity for muddling through in imperfect and messy situations—allows our instincts to create and put up with near intolerable situations. The link between our long- and short-term interests is thus broken. Sometimes our immediate drives override longer-term considerations. The demands of a business for quick bottom-line results, for example, often turn out to be destructive, such as when downsizing to raise shareholder value eats into corporate muscle long after it has finished removing the fat. The reverse is also common. Chief executives' grandiose long-range visions—their pet five-year plans—blind them to immediate threats from both inside and outside the business. In its darkest days of crisis, IBM was packed with visionary planners who were building for an imaginary future that would never arrive. Not just in business but in all areas of society, we allow bad social experiments to go on longer than they should, while valuable institutions, such as the extended family, wither away.

Using the insights of EP, can we reach toward better accommodations with our surroundings? In the twenty-first century, as never before, we are facing new choices about how to live and organize. The miracles of the information revolution are promising to deliver the stuff of science fiction dreams. No area of life will be untouched by the new opportunities. If we let technologists design our future, we could end up in some new painful traps. As it is, people are already suffering in refined new ways because of the IT revolution. Some people feel more oppressed, stressed, and disempowered than they ever did in the bad old days of conventional organization.

I recently talked to a former coordinator of a call center whose job was to monitor the work of three hundred people, mainly women. She described it as totally "soul-destroying" for her, even more than for the people under her surveillance. She quit the job to set up a stress management consultancy!

The EP answer is that we can find ways of working with rather than against the forces of human nature by learning to understand them. This is not "back to the future" conservatism, nor does it mean surrendering helplessly to our impulses. What it does mean is that we can manage our instincts with insight. Recognizing the real nature of sex differences does not mean that women have to let men walk all over them. Rather, it can help us to find the most creative and productive ways of combining men's and women's distinctive needs and contributions.

EP reminds us of the essence of human psychology and the primacy of human instincts and human relationships in working life. Almost every significant event and development in the business world, and most of the problems too, originate in the motives and choices of individuals and groups.

The best things that happen in organizations happen because of something good some person does for, to, or with someone else.

The worst disasters, miseries, and acts of oppression occur for the opposite reason: people treating other people badly.

One of the most significant discoveries of social science in this century has been that the way we design organizations has a decisive impact on whether we bring out the best or worst in our nature. In other words, structure helps shape culture. EP can give us guidance on several key dimensions of the way we structure our business life.

ORGANIZATIONAL STRUCTURE I

Size: How Big Is Beautiful?

MANY ORGANIZATIONS appear to break the Rule of 150. The major utilities, multinational oil companies, the conglomerate family firms of the Far East, and public service organizations the world over have huge armies of employees. At the end of the twentieth century the world's largest employer, Indian National Railways, had 1.6 million employees. The largest business in America was Wal-Mart, which had a total of about half that number. GM came in next at about 600,000.

A consistent stream of research has shown that the psychological climate and workers' physical health are better in smaller

organizations. So why do big businesses exist? The obvious rational answer is economy: the cost-efficiency benefits of greater
scope and scale. The less obvious but equally important reason is
ego: growth is a mark of success. Owners and executives grow big
companies and are often motivated to pursue inadvisable mergers
and acquisitions for simply this reason. Of course, they will offer
more conventional rational explanations for their actions, but in
many cases they are yet another virility display by ambitious men.
Mergers can come unstuck when more than one big ego is
involved. The first attempt at a merger between the two drug
giants Glaxo-Wellcome and SmithKline Beecham fell apart in
1998 because their respective bosses couldn't agree on how to par-
cel out the power.

How, then, do large businesses manage to engage the loyalty
and commitment of their people? They call upon two key mechanisms of human nature: the attraction of extended family–sized
subgroups and identification through symbols.

Some big companies that create strong loyalty among their
employees do so through a degree of self-management at the local
level. In retail businesses such as Wal-Mart a community culture
is created at the store level. In an industrial business, such as a
Korean *chaebol,* there is a strong group spirit in local factories.
But if you are one of three hundred telephone operators on the
floor of a bank call center, what entity has any real meaning for
you? Your "group" is likely to be the group of individuals at work
whom you know and like. This ad hoc psychological linkage is
what management writers used to call "informal" organization. It
was held to be something of a problem because it cut across the
lines of the formal organization. Trade unions drew much of their
strength in the days of the classic machinelike factory from
people's need for a sense of extended family at work.

Can the same effect be created by using symbols and propaganda? We easily adopt symbols that relate to categorizations of
human groups, and we are very susceptible to identifying with

them, especially if they raise our status by their exclusivity or elitist quality. Among business organizations IBM achieved this on a remarkable scale through its corporate uniform and paternalistic values. It linked an elaborate set of norms, extending to personal conduct and dress code, to guaranteed lifetime employment, intensive employee communications, recognition schemes, and a raft of fringe benefits.

The binding power of this culture turned out to have a significant drawback: it looked too much inward and not enough outward. Customers were neglected. Innovation came slowly. Decision makers were resistant to change when a radical shift was most needed. And when the business did start to change, loyalty in several areas turned to fury against the company. IBM had built its culture not through symbols alone but through a policy of focus on its employees, and even then it was not sustainable during the challenges the company faced.

Other companies have tried to use propaganda such as mission statements, renewal seminars, and cascade briefing systems. These are all internal marketing—selling the company as a brand to its own employees as to a customer. Many employees rebel against this and develop counterreactions. Employees can be as fickle as customers toward the business as a brand. Their primary loyalties are going to be channeled through working relationships, which can easily turn people against the company that employs them. In many a retail outlet you can find staff and customers in rebellious alliance against a despised parent business.

ORGANIZATIONAL STRUCTURE II

Decentralization: Empowering Communities

To get employees to truly identify with a big firm, it has somehow to cultivate the feel of a small business. This is inherently difficult because big business requires big bureaucracy to hold it together. In the 1990s many large organizations realized that having massed ranks of telesales staff, machine operators, or audit clerks was not an ideal arrangement for inducing team spirit and motivation—or for securing customer responsiveness.

Many firms decentralized their operations to secure the twin benefits of staff satisfaction and customer service. But since then many have swung back to centralization, for three reasons. One is that IT innovations have made it more possible to exercise remote control from the head office through surveillance and information feedback systems. Second, decentralized companies found cost and efficiency difficult to control, especially cost from duplication of effort and wasteful internal competition. Third, they found that local needs could be supplied faster and more reliably through the global sourcing of a refined product range. Procter & Gamble is one of many multinational firms that have flip-flopped between localized and centralized structures. There is a recurrent conflict between the benefits of rationality/efficiency and community/ effectiveness.

We are all moved by our instincts. We also march to rhythms beyond our control. Economy has a life and a rationality of its own (see Chapter 5) that demands satisfaction. Once a machine is built, it demands to be fed, and if the fuel needed to keep it running is the human spirit, so be it. Although it is true that we have a great deal of choice about how we organize, our choices are not unlimited.

Primatologists studying chimpanzees have noted a relationship between food supply and social structure and it has been theorized that the same relationship might hold true for humans. When the food supply is centralized, the status order of the social group becomes monolithic and hierarchical. When the food supply is dispersed, power becomes decentralized and egalitarian. Is there a parallel here with our centralized organizational life and the more free-ranging life of the hunter-gatherer clan? Machinelike organizations with hierarchical unitary chains of command—such as banks, carmakers, utility companies, and the like—are focused around concentrated supply chains. "Organic" businesses, such as consultancies, educational institutions, and media firms, have looser structures that go along with their diffuse supply chains and multiple client bases.

Even within a single industry, the contrast is apparent. In banking, retail operations have hierarchical command and control cultures to concentrate their efforts on delivering standardized products to a stable domestic customer base. Investment banks, on the other hand, graze in globally dispersed fields and have flat hierarchies and team-based operations.

Some organizations have figured this out: if they can decentralize their structure into community-sized units, they can create a winning edge through the advantages of a commitment-motivated culture. Leaders who have done this have usually been guided largely by their instincts. You may recall, from Chapter 1, that Ricardo Semler, the owner of Semco, reached the Rule of 150 by gut feel, after experimenting with various ways of managing his Brazilian pump-manufacturing business.

Percy Barnevik reached a similar conclusion, maintaining community-sized units while growing the highly successful Swedish-based conglomerate ABB to a position of market dominance. Barnevik's inspiration was his memory of growing up in a small family business. His "multidomestic" strategy was to acquire small manufacturing and service units and give them autonomy. Under his leadership, strictly maintaining a very small head office, he grew the business to 200,000 people dispersed in twenty-five countries. The management structure governing this large number of people contained no more than five layers in any of its 1,300 operating units, whose size averaged just over 40 workers per unit.

Others have done the same. The owner-founder of the Gore-Tex business built the company's success on subunit sizes of less than 200. Richard Branson created the booming Virgin music and media business by initially keeping the size of company units to around 50 people to keep the business friendly, local, and in touch. His practice was to oversee his businesses, many of them headed by his personal friends, as if they were extended kinship groups, and to run his head office like a family home.

ORGANIZATIONAL STRUCTURE III

Hierarchy: Flexing the Muscles

IT IS tempting to see these examples as pointing to the end of hierarchy. There is a lot of talk in management circles about creating delayered and nonhierarchical structures. Don't be fooled by it. There has never been a nonhierarchical organization. Some idealistic communes, from religious orders to the cooperatives of the 1960s and '70s, managed to remove the overt marks of status distinction, but in all of them hierarchy persisted in one implicit form or another. People continued to defer to one another, obtain unequal shares of resources, benefits, or desirable jobs, and exercised the power to make decisions for one another.

In fact, when you strip out layers of a hierarchy people can start to get pretty nervous and uncertain about where they stand. Problems can arise when someone who yesterday was a grade below you, today sits alongside you in the same broad band. We should not be surprised that people respond by trying to reassert status distinctions rather than accept this kind of demotion to a position of equality. The pettiest rivalries are always of the "as if" sibling type—between people who are of nominally the same status and in close proximity to each other. The urge to create distinctions is the stuff of career systems. I have conducted many employee attitude surveys in different kinds of companies, and in all of them, at every level, there is a huge majority who want promotion (generally more than 80 percent)—far more than can reasonably expect to get it. And the desire certainly does not go away in companies that claim to have created a nonhierarchical culture.

One of the chief benefits of a long chain of command is that people's commitment to a business can be sustained by the lure of advancement. Achieving many small promotions may be enough to keep employees motivated for the bulk of their career, until they reach their final career plateau in their mid-forties. Then it's too late for them to jump ship without leaving behind their career-long accumulation of experience and benefits. Career management systems in firms flourish on the principle that "hope makes a good breakfast but a lousy supper."

We have become addicted to hierarchical structures—or rather, we have lost sight of the alternative, the fluid, decentralized status order of the hunter-gatherer clan, which is not nonhierarchical but egalitarian in spirit and function. In this "organic" structure different individuals take precedence according to current demands. Everyone's contribution is recognized and distinctive, yet it is understood that individuals will be especially honored and regarded for what they bring to the community. Two types of this kind of organization are to be found. One is the stand-alone entrepreneurial business, the other the highly decentralized firm of the ABB type.

Neither is compatible with a massive corporate hierarchy. The problem is that the old bureaucratic model, at its best, remains a wonderfully efficient and economically competitive means of doing things. The linear hierarchy with a firmly policed division of labor remains the most efficient way of conducting controlled operations. This kind of control is needed whenever products, quality, or services have to be rigorously standardized. The old hierarchical model will persist wherever we want hamburgers to taste alike, automobiles to be reliable, drugs to be safe and efficacious, nursing care to be skilled, and warfare to be disciplined. The classic organizational structure will therefore remain with us for many decades, though we may be able to soften it significantly.

ORGANIZATIONAL STRUCTURE IV

Matrix Management: A Model for Our Times?

IN RECENT decades "matrix" forms have become popular as alternatives to linear hierarchy. These involve long-term multi-functional teams working under dual authority with a primary link to project leadership, and a secondary line to a function head. Nowadays any business that has two interlocking ways of organizing—for example, a food company that organizes both by region and by product—calls itself a matrix. There is a lot of disenchantment in firms that do this. Employees suffer from the competing pulls of multiple bosses. Decision making becomes overly complicated. Huge amounts of time are wasted in meetings that regulate how the matrix is working. Teams compete destructively with one another. There are constant power struggles among people trying to take control or make a name for themselves. Others seek to dig themselves into protected positions amid the ambiguity of unclear lines of authority. No one takes responsibility for making tough decisions.

The ideas we have looked at suggest that we should not be too surprised at these difficulties. A matrix organization is a breeding ground for creating group perceptions that set people at odds with one another. Employees' primary identification will always be with the group they are closest to physically or psychologically. This will be where their friends are or where they feel are the people in their field who can positively influence their career futures, regardless of the "official" lines on the organization chart. These are usually people in the local regional branch or in their own functional specialty. Then battles follow as the head office keeps trying to ensure that the product and its brand management take precedence over regional and functional tribalism.

This is a far cry from the original matrix idea. The great power of the prototype was discovered in project-oriented groups, where, in the spirit of hunting parties, people with various specialties interrelate with a mix of spontaneity and disciplined commitment to achieve a valued common goal. It has been just such multifunctional teams that have built innovative new airplanes and discovered lifesaving drugs.

Making the matrix model work requires three things. First is that the task or project be supreme. This cannot be enforced by edict. As in any high-performing team, people need to care powerfully about both their own contributions and the common task. Second is the business's culture. The community must value and honor the work of the team. This means, for example, that functional bosses are prepared to "give away" their best people to project teams, rather than clinging to them to sustain their local empires. Third is leadership. There needs to be shared, flexible leadership within the team and a team-minded boss running the division or firm to ensure that the matrix does its work.

ORGANIZATIONAL STRUCTURE V

New Corporate Forms: Virtually Possible?

ADVANCES IN information technology, materials, and communications are revolutionizing many businesses and creating new organizational forms. We are seeing some new alternatives to the traditional and matrix forms. Increasingly one hears talk of "boundaryless," "modular," and "virtual" organizations.

All these do offer refreshing alternatives to the strictures of what we have been used to in recent decades, but beneath the hype are certain unchanging realities.

The so-called boundaryless organization, championed, among others, by Jack Welch of GE, is in reality a flexible team-based concept of working. As we have seen, in some industries this works well, but in others it is inhibited by the demands of technology and hierarchy. To the degree that new technologies are making boundaryless organizations possible, than they offer distinct psychological advantages. Yet in all businesses, including GE, there remain real boundaries of status and function that will not go away because people will not let them. People want to achieve status through advancement and recognition of their distinctive skills, and this perpetuates the traditional functional hierarchy.

"Hollow" and "modular" organizations are new varieties facilitated by the increasing possibilities for outsourcing businesses' work and coordinating it through networks of alliances. Thus Reebok owns no factories, and workers at Dell Computer assemble components not produced anywhere within the company's own boundaries. Toyota has long been a pioneer of this kind of mix-and-match flexible organization. The strategy has proved remarkably successful, but we should not imagine that it has replaced the old organizational forms; it has simply displaced them. All these companies' components are produced by people in factories or offices somewhere. Many of the firms producing the components are small family-owned businesses. In effect, these new forms of organizations as decentralization by another route. Certain points touched on earlier apply here. The outsourcing/alliance model works well to the degree that it can capitalize on local identification and avoid high costs. This has been achieved in many cases and suggests that organization based on these new forms will grow to replace some of the traditional large firms in various fields.

The term *virtual organization* does not have a clear identity.

One model is the firm that comes into existence only when it does something. Rock concerts and some record labels meet this criterion. People have always engaged in ad hoc projects under some kind of loose organization, especially in the media. But is this a viable general model for the future? Impromptu and pragmatic association is something that fits well with our psychology, but it is hard to imagine how it can become a general model for doing more concrete things such as, say, making shoes, erecting buildings, or running a banking operation.

The other kind of virtuality growing around us rapidly is to be found in e-commerce businesses. These are, in reality, profit centers for networks of activities. Many are little more than nodes in a network or coordinating agencies. Again, productive activities have merely been displaced. Traditional working businesses continue to supply and maintain these virtual companies' product base.

Beneath the hype, much remains unchanged. What is new, though, is an increased emphasis on relationships. The new information age has broadened the range of ways in which we can conduct these. But the evolutionary perspective reminds us that working relationships will settle around the contours of the familiar tribalisms of blood, friendship, and presumed similarity. Another welcome feature is that some of these forms make new work roles possible. Telecommuting, for example, allows the boundary between work and nonwork, home and office to be blurred. This would be fine if it were embedded in a community of supportive practice and values, as it was in the world of our ancestors. In today's stratified and bounded world, however, it can be a recipe for isolation, family conflict, and career disadvantage.

There is a health warning here. The Internet—the most revolutionary force of current times—allows us to do lots of things that formerly we had to do out of our homes. This presents a big danger for many people. By fulfilling our appetite for information, people to chat with, and entertainment, the Web can deprive us, without our realizing it, of essential experiences. In the same way that a diet of home-delivered pizzas will deprive you of essential vitamins, failure to get out and meet real people and put some effort into earning your experience runs the risk of depriving you of essential stimulation. This comes from such things as the unprogrammed encounters, difficult situations, pleasant surprises, and reality checks that happen unbidden when one goes out of the front door. Research is beginning to confirm this. Internet junkies are showing signs of depressive illness and loneliness. The more they surf the Net, the more their social involvement declines.

Virtual relationships on chat lines do occasionally end up in happy marriages. Research shows that the persistent human requirement for face-to-face interaction means that new Internet friends almost always create an opportunity to meet at quite an early stage of their relationship. Failure to check reality from time to time by stepping out of one's apartment or going to work creates real threats to maintaining one's psychological health and creating meaningful relationships.

ORGANIZATIONAL CULTURE I

Cultures of Uniformity Versus True Community

THE NEW communications technologies also present another danger: the trap of selective association and cultural cloning. Structures are designed. Cultures grow. The conditions created by the Internet and IT mean that cultures no longer have to grow around people different from each other who have to accommodate one another within a shared space. In Internet relationships people don't have to compromise to get along; they can just go off-line.

The wonder of the World Wide Web is its reach. Connections can be made with just about any type of person anywhere in the world. This is great in many ways, but it also carries a threat. It enables the fanatics, zealots, and perverts of the world to hook up with other people like themselves and so get to feel accepted and normal. Why shouldn't they, you might say? The answer is not so much a moral riposte as a practical one. One of the things that keep us healthy, both psychologically and socially, is having to deal with and come to terms with people who are different from ourselves. Committed life partnerships between individuals have a healthy effect on both parties for this reason. By dealing with someone unlike yourself, one learns about one's own impulses and how to moderate them in the interests of creative partnership. Traditional marriage is both so hard and so rewarding because it forces men and women to appreciate each other.

One of the benefits of any kind of uncontrolled association is the likelihood that people different from one another will mix and meet. In the post–World War II decades, many young middle-class mothers became intensely frustrated during their children's early years, feeling locked into suburban life as homemakers, associating only with people like themselves. Many also discovered great satisfaction from going to work and meeting other kinds of people, even when the work itself was uninspiring.

Our need for community is deep and ancient, and the benefits of true community stem from how it forces the pace of our experience. The key features of what I call true community emphasized in this book are:

1. Its people have no choice about the diversity of the society they encounter,
2. The diversity is normal; i.e., it encompasses the full range of human types.

It is true that even in a clan of, say 150 people, one could choose to spend most time with the 30 or so who matched your demographic profile or manifestly shared your interests, but you would still have to get along with many other people. Getting along means communicating, understanding, compromising, doing deals, collaborating—and sometimes engaging in conflict. The learning that comes from these acts enriches our lives and minds at many levels. If you like, it makes us better managers, since that is what management is all about: dealing with human differences. The best managers are those who are able to relate to and motivate people who differ radically from themselves as much as with those who are similar.

In many contemporary businesses, the trend is in the opposite direction, toward refined specialization. Software houses are full of obsessed technicians, accounting firms full of control addicts, law firms full of game players, and publishing houses full of frustrated writers. This is a caricature, of course, but the problem is real, and it leads to serious organizational pathologies. Here are some examples:

◇ Police forces have extreme difficulty ridding their ranks of bigotry because they attract people with authoritarian attitudes who want to maintain social control.
◇ Science-based businesses have difficulty fostering entrepreneurial leadership because they attract perfectionists and theorists rather than risk takers and people committed to action.
◇ Media firms suffer from an excessive number of creative dreamers—powerful people who like launching initiatives but are reluctant to do the hard grind of seeing them through to completion.

Not every accountant is uncreative. Not every police officer is closed-minded. Luckily diversity cannot be prevented from entering an organization, but its benefits can be suppressed by bad management. The critical issue is what is done with diversity.

What is dangerous is penalizing people who deviate from the norm. If an unartistic recruit to a publishing house not only finds she is different from her colleagues but that they move through the ranks faster than she does, she will quit. Sometimes such people get locked into this kind of misfit by the labor market—there are no equivalent positions to be had elsewhere. If that is the case, such an employee will just continue to sit in her slot feeling unhappy and being less productive than she might be. The tragedy is that someone with her unusual profile might be just what the firm needs. What's more, there is no way she will encourage her like-minded friends to join the firm, even to help solve the problem of her loneliness.

Over time, this sequence of events refines the character of a firm's human capital. Unwittingly (and wittingly), current power holders are selective in favor of their dominant "type," causing the cultural profile of a firm to become more and more narrow. This can seem quite comforting to existing employees. What's more, the firm, if it is in a specialist niche, may do quite well because of this refinement. But such uniformity makes a firm inflexible and nonstimulating for employees and over the long run makes it less well equipped to take control over its destiny in times of change.

ORGANIZATIONAL CULTURE II

Healthy Business As Community

THROUGHOUT THIS BOOK, I have repeatedly raised the prospect of the organization as community as a model of psychological health, environmental adjustment, and business effectiveness. Is this just a folksy idealization? Is the image of bygone clan life a piece of sentimental romanticism? Might not some of the most effective businesses be those where people fight continually to win against all comers? Might not the happiest firms be the most inefficient, full of people just grazing happily with like-minded friends and colleagues?

First, it is true that communities—whether ancestral or otherwise—are not always awash with peace and love. People suffer, fight, and fail in *all* communities. The test of true community is what one does about these experiences. Healthy communities are self-regulating. They solve their problems in ways that do not leave a destructive residue of reduced effectiveness or long-lasting psychological scarring.

It is easy to be deceived by outward appearances. Let's consider the case of what I shall call Company X, a business that seems to have created an effective cultural integration based on principles quite different from those I have espoused. But a closer look reveals that it has done so only at the cost of becoming pathological. Company X is a sales organization; it operates on dog-eat-dog principles. Everyone depends upon his immediate superior's approval for survival. Of course, the superior is also subject to this regime. Failure is punished swiftly and decisively. There is a steady stream of people who enter and exit—fast. Some survive and join the ranks of the others who can tolerate the climate. In fact, some have been there quite a long time. A high percentage of earnings in the firm is commission-based. The boss determines employees' earning opportunities by the client list he (rarely she) feeds them. The people who stay make a lot of money under conditions of zero security. Many people seem very happy with these arrangements, and the company is an outstanding financial success.

How is this possible? The mainstay of this operation is its selection-retention system. People who don't like the environment last barely days. Those who remain feed off one another and the shared culture. A team of MBA students who completed a case report on the firm were horrified. Their outsiders' view was, how could people bear to work in such a place?

Indeed, it is possible for a company like this to be highly

profitable and keep its employees happy without coming close to the communitarian ideal. But is such a business healthy? By that I mean, what are the hazards it creates for its people and its business prospects? The omens are not good. Its fortunes can be bright only in the short term. The talented people it will need to move on to its next stage of development will find the environment uncongenial. Company X is a business that is doubly vulnerable: to breakdown of its internal culture and to failure of its external adaptiveness. It is vulnerable to destructive internal conflicts, and if the market requires a different way of operating it would find the adjustment very difficult to achieve. Currently it is surviving, but I wouldn't invest in such a firm. Would you?

It is my contention that the long-term prospects are best in places that create the values of true community, where people of many kinds can be fulfilled by feeling and being personally effective. Increasingly, evidence is stacking up to confirm this view. We have seen several examples in this book. For more of these, at around the second week in January this year look for the feature in *Fortune* magazine that lists "the 100 Best Companies to Work for in America." Most of these firms are not only places in which people *love* to work, but they are also almost all outstandingly successful in business terms. The same names appear repeatedly: Southwest Airlines, Goldman Sachs, Hewlett-Packard, Federal Express, Marriott International, Johnson & Johnson, Harley-Davidson, Herman Miller, 3M, and Nordstrom. To these big names you can add many small and obscure businesses that are growing fast by using the same formula.

What they share is business drive consolidated around a unified value set. *Respect* is the key. What does this mean? Several things. It means treating customers as honored guests. It means caring for the whole person who works for you, not just the part of that person's identity that relates to work. It means trusting people to act responsibly if you give them the chance. It means celebrating and sharing everyone's achievements. It means cultivating team spirit.

Almost without exception, these companies utilize what we might call *inclusive* features and practices. Almost all emphasize what used to be described as "paternalistic" provisions for their staff: fringe benefits, family-oriented social events, flexible working arrangements, co-ownership shares, and the like. All seem to recognize the simple truth with which we started this book: the value to people's well-being and the effectiveness of companies that give top priority to people-centered systems, values, and practices.

Here's a recipe you might derive from this list of EP principles we have discussed. Not all ingredients are essential to every business, but the more of them that are present, the better:

◇ A unit size that is small enough for all employees to relate to all others.
◇ A flexible hierarchy in which project leadership can change organically.
◇ A climate of recognition for individual contributions.
◇ Team projects as the main modus operandi.
◇ Forums for the free exchange of information, ideas, and feelings.
◇ High involvement of people in the implementation of decisions that affect them.
◇ Norms allowing open expression of emotions.
◇ Self-examination at all levels of successes and setbacks without fear of failure and recrimination.
◇ High-trust systems with near-equal sharing of benefits.
◇ Value-driven leadership emphasizing the value of diversity.

The last point is especially important, since it can propel all the others. A lot was said about leaders earlier in this book. Let us come back to them now, for their stewardship of the business culture is their most important task. They can make or destroy the community's values.

ORGANIZATIONAL CULTURE III

Leaders: Custodians of Culture, Builders of Community

W E CAN design organizational structures, institute systems, publish crystal-clear mission statements, proclaim the most humanistic value systems, and all will be so much wind and brass without a conductor. When it comes to the final performance the conductor might seem superfluous, but his real work, blending the orchestra into a true musical community, lies behind him. In the words of Lao-Tzu,

> As for the best leaders, the people do not notice their existence. The next best the people honor and praise. The next, the people fear, and the next the people hate. When the best leader's work is done, the people say, "We did it ourselves."

The leader's critical role in defining the strategic mission of a business and building relationships to achieve it was discussed earlier. The combination of these two amounts to the creation of a corporate culture, and a leader's special responsibility is to help build and sustain the power of community.

The leader has two means by which to do this: structure and communications. Curiously, the leadership literature has relatively little to say about these. Perhaps the first is neglected because in many organizations structures are taken as given. Writers are more interested in the way leaders manage structure. In my experience this underestimates the power leaders have—or rather the power they can seize if they have the nerve. A new leader's first act is often a structural reorganization, but often this is only a political stratagem—a device for burying certain people and elevating certain others, not a worked-out set of ideas about how to organize to achieve changing business goals. In the 1970s and '80s IBM went through more reorganizations than any other major corporation. It was not until Lou Gerstner radically broke with the past to refocus the business in both its structure *and* its culture that its fortunes began to change. There is a lot a leader can do to change a firm through altering its structure. He or she can:

◇ Set goals
◇ Change how people are grouped
◇ Determine decision-making methods
◇ Establish teams' accountabilities
◇ Fix reporting lines
◇ Calibrate reward systems
◇ Set entry and exit criteria

These and many other things within a leader's grasp can be enormously instrumental in shaping a business's culture.

The leader's communications role is that of cultural custodian, tuning in to the language, rituals, symbols, and meanings that embody the culture of the working community. Cultures are defined by communication—values and beliefs have to be disseminated—and the leader's role is paramount. In a few or many words the leader can capture the essence of collective endeavor.

As we saw in Chapter 4, the most revered bosses view their leadership as a service to the people, a custodianship within the firm's life span, and a trust held on behalf of future as well as current generations. Like a mother or father of a family, a chief executive's interests and goals are transcendent and durable. What matters is that the gene pool he or she leaves is stronger than when he or she arrived, certainly not that he or she accumulated personal riches or was famous for a day or two.

This does not mean that leaders have to be charismatic. They just have to keep faith with the culture, which means helping the emergence of the next generation by letting go early enough. Mismanagement of leadership succession is a fatal indicator of the leader's neglect of the cultural role. A sign of this neglect that has always shocked me more than any other when doing company surveys is how surprised leaders of most medium- to large-sized businesses are about what their own employees think. Until a survey is conducted, many have little or no idea about their employees' true concerns and desires. Even more disturbing is how commonly they drastically underestimate their employees' pride and, loyalty—how much people are willing to give of themselves for the sake of colleagues, customers, and the firm's prosperity. Self-interested political leaders seem to forget the communitarian instincts all too easily.

We have already looked at several of the causes of this deficiency:

◇ Firms become too large for their leaders to get close to their people.

◇ Leaders are surrounded by yes-men and yes-women who shield them from the truth about what is happening in the company.

◇ Systems tie up the leaders' energies in administration and externalities to the exclusion of building community.

All the talk about inspirational leadership misses the point about what leaders should be doing. The cultural mission of leadership demands more than the cosmetics of charisma. It is the embodiment of the Duke Ellington Principle (see Chapter 6)—a vision of what can be achieved collectively, the foresight to create the structures to enable people to make music together, and focused communication to bring out the distinctive sound of the orchestra.

ORGANIZATIONAL CULTURE IV

In the Hall of Magic

IN 1999 I was a guest at a group meeting of the top 250 managers of a pharmaceuticals company. The event breached the Rule of 150 in its scale, and clearly it was not a forum for all these senior managers to get to know each other, but it was manifestly and latently an important cultural event from start to finish. It started with all the executives boarding a specially commissioned train to take them to the conference location. On the train, hired actors and actresses acted out sketches with the executives around the mission of the company. During the conference there were various other cultural events: speeches, award ceremonies, workshops dealing with strategic themes. There was plentiful opportunity for gossip. The concluding dinner of the conference was celebrated in the big top of a full-blown circus, officiated by one of the main directors acting as MC. At one point in the event, as a symbol of the company's commitment to lifelong learning, everyone was given juggling balls and taught how to juggle by the circus clowns. Even people who normally hate "this sort of thing" told me they got caught up in the spirit of the occasion and had a lot of fun. Here was a family firm trying to involve its people— albeit its topmost layer—through symbols and theater.

Increasingly, one can see other examples of similar events in business: managers attending jazz improvisation workshops, executives doing role plays with drama professionals, leaders creating visual arts in groups to "liberate the right side of their brains." A lot of nonsense theorizing will accompany these activities, and one should not take too seriously what people say about what they are doing. Their most important aspects are to fulfill cultural needs and expression.

No doubt the wage slaves at the bottom levels of the business have a cynical view about what their executives are doing at such events. But that is because they are excluded, and indeed one danger of top management culture-building activities is that they widen the fault lines between the elite and the rest—perhaps especially those who just fail to make the invitation list. Smart businesses leaders recognize that everyone needs magic.

Let's just define what we mean.

The magic that is practiced in tribal communities is a far cry from the stage trickery of the developed world. It focuses on the key events of the life cycle of individuals and the tribe: birth, coming-of-age, marriage, death, the accession of leaders, hunting, the time of year, weather changes, sickness, and health. Magic and ritual are not so much supernatural as natural. Rather than trying to create something out of the ordinary, the shamans, witch doctors, and priests of the tribe bring to life the connection between people's native psychology—their hopes, fears, and dreams—and the world of vital experience. For example, hunter-gatherers often revere as sacred the beasts on which they most depend for their sustenance. Magical stories represent the relationship, and the killing and consumption of these animals are heavily ritualized.

The cave art and elaborate tool decoration of Stone Age peoples represent all the important events and relationships of community life. There has been a tendency to explain away these forms as the attempts of primitive peoples to "make sense" of a world they could not understand scientifically as we do now. This is looking at the past through the rationalistic lens of our present time. Magic and ritual are more deeply instinctive than this and part of our hardwiring. It is plain that we too as modern peoples retain a yearning for magic, an enchantment with art, and an addiction to ritual. As fast as Western societies abandon their traditional faiths, up spring New Age and millennial fantasies, from UFOs to faith healing. Business is full of invented ritual, from the selection process to the exit interview, many of them managed by the Human Resources Department—but without much magic, it has to be said!

Our need for magic comes from our need for confidence and control over the dangers and uncertainty that surround us. We have seen this throughout this book, starting with my reflections on my irrational thoughts concerning my mugging (see Introduction) to the magnetic attraction bad news holds for us.

Our minds are constructed to respond warmly to symbols, art, rituals, and other displays. They engage the powerful capacities of our visual systems, our categorical and linguistic intelligence, and our emotional responses. It is important for us to feel connected with the natural world that surrounds us. Instead, we increasingly find ourselves cut off from our environment and even our own human nature by the conceit of our rationality and the systems that elevate it.

The nearest one gets to magical ritual in some companies is an embarrassed gold-watch retirement party, so perhaps we should be grateful that companies are beginning to loosen up by using more playful training methods. But this is insubstantial stuff. In smaller firms—less than 150 strong—it is possible to have more inclusive events, such as "town hall" meetings where everyone gathers to discuss the fate of the firm or whatever is on people's minds. Jack Welch tried, with some success, to cultivate a spirit close to revivalist prayer meetings with his "workouts"—occasions when he would conduct open question-and-answer sessions with groups of workers all over GE's gigantic empire. Bill Gates opened himself up to e-mail dialogue with all Microsoft's employees. These are leaders trying to do cultural work. Let no one be fooled that these are always effective ways of solving problems or efficient ways of fixing things. They are far too random, unsystematic, and, frankly, ostentatious. But they do succeed in creating positive beliefs. Companies that do this well are the more powerful for it.

ORGANIZATIONAL SOCIETY
IN THE FUTURE

CAN WE already see the future of organizational society in places near at hand? In Silicon Valley, for instance? Certainly the firms the Valley hosts do utilize many radically new ways of communicating and organizing, but we should beware of imagining that this is completely novel. In Silicon Valley people have jobs and careers, albeit in a networked landscape. Many of its organizations are still housed in physical buildings. People do chat by electronic media, but lots of face-to-face meetings go on all over the place.

What is so impressive about Silicon Valley is its vibrancy as a community of invention and entrepreneurship. This is due to the *physical* clustering of similar businesses geographically. Economists and professors of strategy have long remarked on the importance of industrial clustering and the curious way businesses seem to benefit from being geographically close to their competitors, rather than securely ensconced in some faraway niche— much as runners record their best times against real competition in a race rather than alone against the clock. Clustering brings other advantages: hosts of small suppliers, intermediaries, and various others that benefit from being close to the cluster, feeding it and feeding from it. This forms the basis of a tribelike agglomeration of interlocking quasi-kinship groups of people who can circulate and innovate as they move around.

This pattern, though highly dynamic, is not new. In the late nineteenth century, firms clustered together and innovated in much the same way—the textile mills of northern England, for instance, or the engineering firms and carmakers of the North

American Midwest. Just as today's software programmers cluster around the Valley, so did innovators cluster around the industries of earlier times.

Silicon Valley is held up as a modern marvel for its astonishing rates of innovation, inspired by its mobile population of talented workers. It resembles a cross between the gold rush to the Yukon in the 1890s and the Hollywood dream factory of the 1940s. It is a magnet for talent. But, like Hollywood, it is also a graveyard of broken dreams. Under the shiny surface of Silicon Valley is an unattractive substratum where much is secret, competition is ruthless, and weakness and failure are punished without mercy. The Valley can exist only on the basis of refined selection—its ability to import the talent it wants and discard the talent it has used up. This makes it dependent on a wider community on which it feeds. For this reason the whole world cannot be like Silicon Valley: there can't be a top of the heap without a pile beneath it.

We all benefit from the results, and as long as we do, clusters such as Silicon Valley will continue to exist. Is the Valley a model of the future of business communities? In a way it is, underlining how, when other constraints are removed, communitarian instincts can drive the way we organize. In Cambridge, England, Bangalore, India, and Hong Kong, China, biotechnology, software, and electronics companies are already flourishing. New business clusters will continue to arise in new places as long as we find new things to produce and consume.

More widely, as we enter the age of biology—for surely the science of genetics will be a dominant force of innovation and social change in the twenty-first century—we are facing challenges that really are unprecedented. Through genetic engineering we have the power, for the first time, to control key aspects of our own evolution. Very soon we will have the capacity to apply genetic tests that can reliably indicate which aspects of individuality will advance or prejudice people's life and work chances.

This raises huge ethical and practical questions, many of which will involve businesses and their leaders. It is imperative that we understand and take note of the hazards inherent in human nature lest we destroy our world through the kind of arrogant meddling that comes so easily to us.

A NEW WORLD OF WORK?

WE ARE surrounded by soothsayers and futurologists who foresee an unrecognizably transformed world in which jobs, careers, and organizations as we know them will no longer exist. It is true that superficially the world will be transformed. Electronic paper will replace the kind of material you are holding in your hands now. Communications and entertainment will be altered radically. Some areas of work will be unrecognizable. For example, advances in telephony rapidly created the call center as an employment phenomenon of the late twentieth century, and just as quickly advances in voice recognition will make the factory-like call centers of today disappear again within a short space of time.

All this reflects an increasing turbulence and a rising pace of change in patterns of employment, not a revolution in the nature of work itself. As I have argued, the core of business will remain the same. Jobs will continue to exist. Organizations will continue to be located in buildings. People will continue to have careers.

Only in the realm of science fiction will we ever use instantaneous matter transporters. No virtual yachting experience will substitute for sailing with the salt spray in your face. Never will an electronic greeting feel like a handshake or a hug. No digitalized electronic chat room will simulate the warmth of conversing about ideas with friends over a bottle of wine.

So before we get overexcited about the new world opening

before our eyes—and there *is* plenty to marvel at—let us keep at least one foot on the ground. The evolutionary perspective reminds us about what people are likely to continue to want and what they won't put up with.

Here are my seven predictions about what will *not* be altered in the future world of organization:

1. We can adapt quite well to a semi-itinerant, mobile lifestyle. It takes us close to our origins, but it will be popular only where it is embedded in the context of a supportive surrounding community. Because this is not the norm in our culture, only a minority will opt for this type of work and lifestyle. Most people will retain a strong preference for more traditional forms of employment.

2. The traditional idea of the career is a modern invention to fit the linear hierarchy of organizations and occupations. But even if these structures disappear, people will still aspire to lifelong status advancement. If for no other reason, hierarchical career systems will be retained throughout the world of work. Some people will enjoy what are now being called "portfolio" careers, but those who can will stick with what they know best and like to do.

3. The separating boundary between work and home is also an artifice of recent times. We can well adapt to more integrated ways of living, for example through telecommuting, but this too will require a supportive context. People will prefer traditional working patterns if home working dislocates them from their employment community and takes them away from career opportunities.

4. So-called virtual organizations will grow, in the sense that electronic linkages will become predominant modus operandi. But they will quickly succumb to the human instinct to organize hierarchically and clannishly. Networks of advantage, trust, and alliance will govern their conduct. They

will not be preferred employers for the many people who want more direct experience of community.

5. People will always want to make and have "things." We cannot eat virtual bread and drink virtual water or fix cars with virtual tools. Much business will remain in highly traditional forms to make things we like and need to have. Forecasts of "the end of work" and the "death of jobs" are wild and false exaggerations. We will retain much that is familiar about occupational life.

6. People will always desire face-to-face dealings. No amount of virtuality will replace the irreducible experience of physically being in the presence of other people, especially in small groups. From boards of directors to project teams, meetings will continue to be conducted in this manner, even when there are highly sophisticated alternatives.

7. People will always want to congregate in common spaces. They will continue to want to go to work to see and experience the presence of other people. For the same reason that VCRs did not put movie theaters out of business and sports stadiums continue to fill up even when a game can be watched better on TV at home, people in business will find reasons to meet and gather physically.

EVOLUTIONARY PSYCHOLOGY AND CHANGE

THE HISTORY of postagrarian civilization has been largely a story of humans reacting with limited insight to changes they wittingly or unwittingly engineered. We have become much more knowledgeable, but the rate of change has become faster and life infinitely more complex. We continue to be outpaced by our own inventions.

The answer is not to slow down our rate of innovation. The

push for progress is an ineradicable human instinct. It is not an option for us to return to our Stone Age way of life, but we can understand what living and working in harmony with human nature means. The EP analysis suggests that this means we need not more differentiated sex roles but more equal representation. We want less dominant and more flexible leadership. We need less formal hierarchy and more community.

The only other way of coping with the pace of change and level of complexity is through insight. EP also points firmly in the direction of ourselves and our unawareness as the chief obstacle to adjustment. If we can understand a bit more clearly the essence of what we are, we can design situations to fit our impulses, get advance warning about our own reactions, and in some cases redirect our energies. Just because we have instincts doesn't mean we don't have choices.

The main problem is deciding *what* we want. Perhaps we have too many choices. Or to put it another way, every situation we can engineer has a downside of maladjustment as well as an upside of attraction. So the question is, what kind of downside is least awful for us? What are we prepared to put up with? In the immortal words of the Delphic Oracle, there is only one answer: "Know thyself." Here is a universal principle I apply in all my work:

> *The system that knows itself has more*
> *control over its own destiny.*

The human psychological system can function better through self-knowledge. All psychotherapy operates on this principle. The best organizational consulting does, too. Organizational systems that comprehend their own processes can operate more effectively. All the best teaching, leadership, and teamwork draw upon this principle.

Evolutionary psychology adds to our insight. It can help

humans liberate themselves—not by removing their impulses but by managing them with insight. EP can help us consider what goals are realistic, and our imaginations can supply the vision. Being human is dangerous. We can do great harm to ourselves and the beautiful world whose creatures we are, just by giving careless rein to our impulses. We can also use our impulses to do great good. Through intelligent organization and management, we can create and sustain the best of human community. For the sake of future generations, we must.

Notes

INTRODUCTION: INSTINCT IN ACTION

PAGE

1   *Managing the Human Animal:* For a short summary of some of the key ideas in this book, see my article "How Hardwired Is Human Behavior?" (*Harvard Business Review*, July–August 1998, pp. 134–147).

2   *We stopped evolving:* The central argument of this book—that our Stone Age nature remains unchanged—does not exclude the possibility of variations on the human design (e.g., immunities to certain diseases, skin pigmentation) being accentuated at particular times and places through the continuing influence of natural and sexual selection. The point is that nothing has or could have reshaped our mental architecture from its hunter-gatherer origins. For a general introduction to the field, see Robert Wright, *The Moral Animal: The New Science of Evolutionary Psychology* (Pantheon, 1994) and David Buss, *Evolutionary Psychology: The New Science of Mind* (Allyn and Bacon, 1999).

3   *It was six-thirty at night:* This is the first of many stories in this book. Apart from those about specific businesses and personalities, all other examples are culled from experience, with identifying features scrambled so that any resemblance between them and real-life cases is purely coincidental.

4   *Urban story:* On the morning after this mugging episode I had a meeting with David Cannon, at that time my Ph.D. student, who was studying psychological reactions to failure. We discussed the event at length, and his insights brought home to me the forces that create such emotional reactions and magical thinking as ways of gaining control over uncertainty and danger. These and other thoughts from conversations with David have been a major influence on the development of this book. See his article "Cause or Control? The Temporal Dimension in Failure Sense-Making" (*Journal of Applied Behavioral Science*, vol. 35, 1999, pp. 416–438).

13   *EP's insights do not eliminate choice:* Various fallacies in popular conceptions of evolutionary psychology are discussed by Livia Markóczy and Jeff Goldberg in "Management, Organization and Human Nature: An Introduction" (*Journal of Managerial Economics*, vol. 19, 1998, pp. 387–410).

1. STONE AGE MINDS IN THE INFORMATION AGE

20   *The Ancestral Environment:* Numerous books trace the evolution of the journey from the first bipedal hominids to modern humans. See, for example, Christopher Wills, *The Runaway Brain* (HarperCollins, 1994), and R. Boyd and J. B. Silk, *How Humans Evolved* (Norton, 1997).

21   *What we were then:* We cannot know for sure what it was like, but reasonable inferences can be made from observations of surviving hunter-gatherer peoples, archaeological remains, and inference from human design. Numerous books give more detailed accounts of this ancestral environment. See, for example, Tim Megarry, *Society in Prehistory* (Mac-millan, 1995).

22   *Since we have no way:* Scholarly reconstructions of our ancestral environment and how it shaped our psychology can be found in M. Corballis and S. E. G. Lea, eds., *The Evolution of the Hominid Mind* (Oxford University Press, 1998), and Paul Mellars and Kathleen Gibson, eds., *Modelling the Early Hominid Mind* (McDonald Institute Monographs, 1996). See also Marek Kohn, *As We Know It: Coming to Terms with an Evolved Mind* (Granta, 1999), and Craig Stamford, *The Hunting Apes* (Princeton University Press, 1999).

22   *Darwin's Framework:* There are numerous accounts of Darwin's theory and its current status. One of the most influential has been Richard Dawkins, *The Selfish Gene* (Oxford University Press, 1976). A very thorough account is also to be found in Daniel Dennett, *Darwin's Dangerous Idea* (Simon & Schuster, 1995). A good discussion of issues in contemporary genetics can be found in Steve Jones, *The Language of the Genes* (HarperCollins, 1993).

23   *But new genes have to do more:* Probably the most complete scholarly account of the key idea of sexual selection is in Helena Cronin, *The Ant and Peacock* (Cambridge University Press, 1991). See also Matt Ridley, *The Red Queen* (Viking, 1993), and Geoffrey Miller, *The Mating Mind* (Heinemann, 2000). The last of these shows how the theory explains many phenomena in everyday life.

23   *It is now reckoned:* This phenomenon is called self-handicapping. See A. Zahavi and A. Zahavi, *The Handicap Principle: A Missing Piece of Darwin's Puzzle* (Oxford University Press, 1997).

24   *The Neuropsychology Revolution:* Neuroscience is developing at a rate that challenges the bibliographer. See Ronald Kotulak, *Inside the Brain* (Andrews & McMeel, 1995), and Rita Carter, *Mapping the Mind* (Weidenfeld and Nicolson, 1998). For a discussion of brain functioning from an evolutionary perspective, see Stephen Pinker, *How the Mind Works* (Norton, 1997).

25   *Genes for individual abilities:* An evolutionary account of the genetics of health and illness is provided in Randolph Nesse and George Williams, *Why We Get Sick* (Times Books, 1994).

25   *Archaeologists join the hunt:* An archaeologist's reconstruction of

human evolution is offered in Steve Mithen, *The Prehistory of the Mind* (Thames & Hudson, 1996).

26 *A key conclusion:* We are still discovering the exact historical sequence and environmental causes that led humans through the crucial stages of migrating out of Africa and subsequently adopting the lifestyle of settled agrarians. Works on this include Jared Diamond, *The Rise and Fall of the Third Chimpanzee* (Random House, 1991), and Colin Tudge, *Neanderthals, Bandits and Farmers* (Weidenfeld & Nicolson, 1998).

26 *These are the conditions:* Evolution proceeds rapidly and then stabilizes so that species generally do not evolve in their own lifetime. This is the idea of punctuated equilibrium. See Nils Eldredge, *Reinventing Darwin* (Wiley, 1995).

26 *What has changed for humans:* Taking a lead from an idea in Dawkins's *Selfish Gene*, the idea of a science of "memetics" has been suggested by Susan Blackmore, *The Meme Machine* (Oxford University Press, 1999). This proposes that cultural ideas can evolve and reproduce within human minds. A less controversial and speculative account of the biological basis of cultural evolution is provided by Dan Sperber in *Explaining Culture* (Blackwell, 1996).

27 *Thus was modern society born:* Diamond also describes how geographical dispersal affected the uneven march of civilization in *Guns, Germs, and Steel: The Fate of Human Societies* (Norton, 1997).

33 *Management as Nature Intended:* Semler gives a full account of his experiments at Semco, from the time he took over the firm up to his partial withdrawal into the guru lecture circuit, in his book *Maverick!* (Arrow Books, 1993).

35 *"What if I could strip away":* Ibid., p. 65.

2. MANAGING AGAINST THE GRAIN OF HUMAN NATURE

39 *Chapter 2:* An extended discussion of the phenomena described in this chapter can be found in my article "The Seven Deadly Syndromes of Management and Organization: The View from Evolutionary Psychology" (*Managerial and Decision Economics*, vol. 19, 1998, pp. 411–426).

41 *Remnants of the First Age:* An account of recently surviving hunter-gather peoples can be found in Carleton S. Coon's *The Hunting Peoples* (Cape, 1972).

41 *This is happening in pockets:* The oppressive underbelly of life in Silicon Valley is vividly described by Po Bronson in *The Nudist on the Late Shift: And Other True Tales of Silicon Valley* (Secker & Warburg, 1999).

42 *As we enter the Fourth Age:* Evolutionary psychiatry is emerging as a new Darwinian way of understanding the incidence of modern mental illness. See Anthony Stevens and John Price, *Evolutionary Psychiatry* (Routledge, 1996).

44   *Free from the emotional burden:* See *Shame and Related Emotions: An Interdisciplinary Approach,* special issue of *American Behavioral Scientist* (T. J. Scheff, ed., vol. 38, no. 8, 1995).

44   *Daniel Goleman:* See his *Emotional Intelligence* (Bantam Books, 1995) and the idea applied to business in *Emotional Intelligence at Work* (Bantam Books, 1998). The neuropsychology of emotions is discussed in Joseph Le Doux, *The Emotional Brain* (Touchstone Books, 1998) and Antonio R. Damasio, *Descartes' Error* (Putnam, 1994).

47   *Quite apart from:* See Deborah A. Waldron, "Status in Organizations: Where Evolutionary Theory Ranks" (*Journal of Managerial Economics,* vol. 19, 1998, pp. 505–520).

47   *Humans are in exactly:* The disorders resulting from low rank have been documented in a huge epidemiological study of the U.K. civil service—see, e.g., E. J. Brunner, "Stress and the Biology of Inequality" (*British Medical Journal,* vol. 314, 1997, pp. 1472–1476). A complete review of the effects of social stratification from an evolutionary perspective is provided in the two volumes of Lee Ellis, ed., *Social Stratification and Socioeconomic Inequality* (Praeger, 1993 and 1994). See also Richard Wilkinson, *Unhealthy Societies* (Routledge, 1996).

49   *Second is sensitivity:* The first thorough evolutionary account of the issue of contract violation and its importance to humans is to be found in Leda Cosmides and John Tooby, "Cognitive Adaptations and Social Exchange" in Jerome K. Barkow et al., eds., *The Adapted Mind* (Oxford University Press, 1992, pp. 163–228). This landmark collection of essays helped to launch evolutionary psychology as a discipline.

49   *In business, people are often unhappy:* There is a large literature on distributive and procedural justice in business. See, for example, Jerald Greenburg, "Looking Fair vs. Being Fair: Managing Impressions of Organizational Justice" (*Research in Organizational Behavior,* vol. 12, 1990, pp. 111–157).

52   *Now get the two groups:* The ease with which one can arouse these feelings is called *minimal group identification.* See Henri Tajfel and Michael Billig, "Familiarity and Categorization in Intergroup Behavior" (*Journal of Experimental Social Psychology,* vol. 10, 1974, pp. 159–170).

53   *In 1997:* Case reported by Janice Warman "What's Good for Tokyo Is Good for You," in *The Observer,* February 2, 1997.

55   *These examples show:* See Denis Collins, "The Ethical Superiority and Inevitability of Participatory Management As an Organizational System" (*Organization Science,* vol. 8, 1997, pp. 489–507).

56   *Groups have to focus:* The requirements for productive group dynamism, free from these rigidities, are documented Serge Moscovici and Willem Doise, *Conflict and Consensus* (Sage, 1994).

58   *These are human biases:* The biases in the beliefs of career-mobile people are described in my book with Michael West, *Managerial Job Change* (Cambridge University Press, 1988). See also Scott E. Seibert et al., "Proactive

Personality and Career Success" (*Journal of Applied Psychology*, vol. 84, 1999, pp. 416–427).

58   *Migrants' optimism:* See J. Stewart Black and Hal B. Gregersen, *So You're Coming Home* (Global Business Publishers, 1999).

59   *The business literature is littered:* There are numerous popular and scholarly accounts of such human errors and disasters. See, for example, Normal Dixon's *On the Psychology of Military Incompetence* (Cape, 1976), and Charles Perrow, *Normal Accidents* (Basic Books, 1984).

59   *We have inbuilt:* A range of biases affecting managers is analyzed in Max Bazerman, *Judgment in Managerial Decision Making* (Wiley, 1994).

60   *Management by fear:* See R. A. Giacolone and J. Greenberg, eds., *Antisocial Behavior in Organizations* (Sage, 1997).

61   *In firms it is:* See Ben W. Schneider, "The People Make the Place" (*Personnel Psychology*, vol. 40, 1987, pp. 437–453).

61   *Sometimes human dominance resembles:* See Richard Wrangham and Dale Peterson. *Demonic Males: Apes and the Origins of Human Violence* (Houghton Mifflin, 1990).

62   *Violence is close to:* See Roy F. Baumeister et al., "Relation of Threatened Egotism to Violence and aggression: The Dark Side of High Self-Esteem" (*Psychological Review*, vol. 103, 1996, pp. 5–33). Violent and aggressive behavior at work (almost exclusively perpetrated by men) has attracted much scholarly attention in recent years. See, for example, Duncan Chappell and Vittorio Di Martino, *Violence at Work* (ILO, 1998), and Ricky Griffin, ed., *Dysfunctional Behavior at Work, Part A: Violent and Deviant Behavior in Organizations* (JAI Press, 1998).

3. SEX AND GENDER

67   *The way we work:* Evidence for the character of our hunter-gatherer clan life has been discussed by Andrew Whiten in "The Evolution of Deep Social Mind in Humans," in M. Corballis and S. E. G. Lea, eds., *The Evolution of the Hominid Mind* (Oxford University Press, 1998). In relation to sex differences, an excellent summary of evidence is presented in Matt Ridley, *The Red Queen* (Viking, 1993).

67   *Yet at the same time:* See Sara B. Hrdy, "Raising Darwin's Consciousness: Female Sexuality and the Pre-Hominid Origins of Patriarchy" (*Human Nature*, vol. 8, 1997, pp. 1–50).

71   *Like the other apes:* How humans conform to the ape pattern of female exogamy is analyzed by Robert Foley in *Another Unique Species* (Longman, 1987).

72   *Genetic differences:* The biology of these genetic effects is described in R. Ohlsson, K. Hall, and M. Ritzen, eds., *Genomic Imprinting: Causes and*

Consequences (Cambridge University Press, 1995). Other biological sex differences are summarized in Anne Moir and David Jessel, *Brainsex: The Real Difference Between Men and Women* (Michael Joseph, 1989).

73   *Biochemical differences:* For an authoritative review of the neuropsychology of emotion and how the male and female brain structures differ, see Ross Buck, "The Biological Affect: A Typology" (*Psychological Review,* vol. 106, 1997 301–336). He shows, for example, how the hemispheres of the brain are wired differently for prosocial and selfish emotions and that men and women differ markedly in the size of these structures. The differences extend to perception and cognition. In 1998 a brain-mapping conference in Düsseldorf showed that completely different circuitries govern spatial mapping in men and women. See also Irwin Silverman and Marion Eals, "Sex Differences in Spatial Abilities," in Jerome K. Barkow et al., *The Adapted Mind* (Oxford University Press, 1992 (pp. 533–554).

74   *On almost any characteristic:* Discussed at length by David C. Geary in *Male, Female: The Evolution of Human Sex Differences* (American Psychological Association, 1998).

74   *Sexual Shopping:* The different reproductive strategies of men and women, and their psychological counterparts, are fully described in David Buss, *The Evolution of Desire: Strategies of Human Mating* (Basic Books, 1994), and Robert Wright, *The Moral Animal* (Pantheon, 1994).

75   *Research has shown:* On the universals of attraction, see Nancy Etcoff, *Survival of the Prettiest: The Science of Beauty* (Little, Brown, 1999).

77   *Yet a similar drama:* See Donald Brown, *Cultural Universals* (McGraw-Hill, 1991), in which an anthropologist provides an inventory of cultural universals, many of them to do with sex differences and gender mores.

78   *Adaptations are always:* See David M. Buss et al., "Adaptations, Exaptations and Spandrels" (*American Psychologist,* vol. 53, 1998, pp. 533–548).

80   *Take, for example:* See Deborah Tannen, *You Just Don't Understand: Talk Between the Sexes* (Morrow, 1990).

82   *Men also do more of the worst-paying:* See Warren Farrell, *The Myth of Male Power: Why Men Are the Disposable Sex* (Simon & Schuster, 1993).

83   *In 1999:* Elizabeth Becker, "Motherhood Deters Women from Army's Highest Ranks" (*The Wall Street Journal,* November 29, 1999).

84   *The Gender Balance Effect:* The process of gender segregation is analyzed in Eleanor Maccoby, *The Two Sexes: Growing Up Apart, Coming Together* (Harvard University Press, 1998). The problems of acceptance of females in male-dominated cultures are discussed in J. G. Berhard et al., *Staying Human in the Organization: Our Biological Heritage and the Workplace* (Praeger, 1992).

85   *A similarly tough:* The story of Susan Estes is told by Gregory Zucker-

man, "How a Woman on a Bond Desk Deals with the Slights and Rises to the Top," *The Wall Street Journal,* November 26, 1999.

86 *Offices with roughly equal numbers:* The reduction of gender relations problems when men and women work together in equal numbers is documented in Barbara Gutek's survey *Sex and the Workplace* (Jossey-Bass, 1985).

89 *Choice of Roles:* See Kingsley Browne, *Divided Labours: An Evolutionary View of Women at Work* (Weidenfeld & Nicolson, 1998).

90 *But these differ:* See "For Better, for Worse: A Survey of Women and Work" (*The Economist,* July 1998).

91 *In the words of:* Numerous comments of this kind were made by informants in our London Business School research project on traders in the City of London. See N. Nicholson et al., "Individual and Contextual Influences on Market Behaviour," end-of-grant report to the ESRC (No. L211252056).

92 *In fact, there is:* The overtrading of thrill-seeking men and superior performance of women's more selective trading was reported in Ruth Simon, *The Wall Street Journal,* "Women Outdo Men in Results In Investing" (October 20, 1998).

92 *Elsewhere, sexual politics:* Sexual politics is the theme of several chapters in J. Hearn et al., eds., *The Sexuality of Organizations* (Sage, 1989). A classic account of gender relations in secretarial roles is provided by Rosabeth Moss Kanter in *Men and Women of the Corporation* (Basic Books, 1977).

92 *Women are also interested:* See Judy B. Rosener, "Ways Women Lead" (*Harvard Business Review,* November–December 1990, pp. 119– 225), S. Helgesen, *The Female Advantage: Women's Ways of Leadership* (Doubleday, 1990), and Alice H. Eagly and Blair T. Johnson, "Gender and Leadership Style: A Meta-analysis" (*Psychological Bulletin,* vol. 108, 1990 pp. 233–256).

93 *The other is to quit:* See Dorothy Moore, *Women Entrepreneurs: Moving Beyond the Glass Ceiling* (Sage, 1997). In her study of American women business owners she found that women reject the corporate environment not because of open discrimination but because they feel they do not belong.

93 *"I am mystified":* Roddick's quote comes from an article by Joanne Martin and others, "An Alternative to Bureaucratic Impersonality and Emotional Labor: Bounded Emotionality at The Body Shop" (*Administrative Science Quarterly,* vol. 43, 1998, p. 447).

94 *Mary Kay Ash:* Her story is summarized in Daniel Gross, *Forbes Greatest Business Stories of All Time* (Wiley, 1996).

4. THE NATURAL SELECTION OF LEADERSHIP

97 *We have entered:* See Scott Adams, *The Dilbert Principle* (Harper-Collins, 1996).

98   *In human society:* See Michael Useem, *The Leadership Moment* (Times Books, 1998), for an exploration of the psychological qualities that define leadership impact. See also Howard Gardner, *Leading Minds* (HarperCollins, 1995).

99   *Great benefits to health:* See Lee Ellis, ed. *Social Stratification and Social Inequality* (Praeger, 1993), p. 49.

99   *We search for heroic:* See Jim Meindl, "On Leadership: An Alternative to the Conventional Wisdom" (*Research in Organizational Behavior,* vol. 12, 1990), pp. 159–204. The prototypical leadership myths are explored in Joseph Campbell, *The Hero with a Thousand Faces* (HarperCollins, 1993; originally published in Princeton University Press, 1949).

100   *Some leaders are not:* See Jerzy N. Kosinski, *Being There* (Corgi, 1978).

100   *Dominance is a hairsbreadth:* Manfred Kets de Vries has written extensively about the dark side of leadership; see, for example, *Leaders, Fools and Imposters: Essays on the Psychology of Leadership* (Jossey-Bass, 1993). See also Baumeister et al., "Relation of Threatened Egotism to Violence and Aggression," *Psychological Review,* vol. 103, 1996, pp. 5–33.

101   *Chris Argyris:* Personal communication, July 9, 1998.

102   *So what happened:* See David Greising, *I'd Like to Buy the World a Coke: The Life and Leadership of Roberto Goizueta* (Wiley, 1997).

102   *Leaders are driven:* My own research on leadership personality profiles bears out the importance of these motives; see N. Nicholson, "Personality and Entrepreneurial Leadership: A Study of the Heads of the UK's Most Successful Independent Companies" (*European Management Journal,* vol. 16, 1998, pp. 529–539).

104   *This can occur when:* The term "machine bureaucracy" was coined by Henry Mintzberg in *The Structuring of Organizations* (Prentice Hall, 1979).

104   *In business, as in politics:* See John A. Byrne, *Chainsaw* (HarperBusiness, 1999).

105   *Even his own adult children:* John Byrne reported on the children's reactions in "How Al Dunlap Self-Destructed" (*BusinessWeek,* July 6, 1998), and in his book, *Chainsaw* (HarperBusiness, 1999).

106   *Bill Gates's obsession:* See Wendy Goldman, *The Microsoft File* (Times Books, 1998).

106   *John Gutfreund:* See Michael Useem, op. cit., Chap. 7, "John Gutfreund Loses Salomon Inc."

107   *The management literature is:* An exception for its close attention to the body language and speech patterns of leaders is Robert J. Richardson and S. Katherine Thayer, *The Charisma Factor* (Prentice Hall, 1993).

108   *Charismatic leaders seek:* Charisma as a function of follower character-

istics and needs is a theme in the work of Jay Conger; see *Charismatic Leader: Behind the Mystique of Exceptional Leadership* (Jossey-Bass, 1989).

109   *Why Leaders Are Born:* The findings of behavior genetics research are presented in Robert Plomin, *Genetics and Experience: The Interplay Between Nature and Nurture* (Sage, 1994), Thomas J. Bouchard, "Genetic Influence on Mental Abilities, Personality, Vocational Interests and Work Attitudes," in C. L. Cooper and I. T. Robertson, eds., *International Review of Industrial and Organizational Psychology* (vol. 12, 1997, pp. 373–396), and David Sloan Wilson, "Adaptive Genetic Variation and Human Evolutionary Psychology" (*Ethology and Sociobiology,* vol. 15, 1994, pp. 219–235).

110   *This sounds contradictory:* See D. T. Lykken et al., "Emergenesis: Genetic Traits That May Not Run in Families" (*American Psychologist,* vol. 47, 1992, pp. 1565–1577). Twin studies use complex modeling techniques to estimate the mapping of shared characteristics in relation to shared genes and shared environments.

111   *Even so, some single genes:* The discovery of dopamine-receptor genes related to personality is reported by C. R. Cloninger and others in "Mapping Genes for Human Personality" (*Nature Genetics,* vol. 12, 1996, pp. 3–4). See also Dean Hamer and Peter Copeland, *Living with Our Genes: Why They Matter More Than You Think* (Macmillan, 1999).

112   *So where does the other 50 percent:* Judith Rich Harris has recently claimed that peer-group influences in childhood explain much of the missing variance; see *The Nurture Assumption: Why Children Turn Out the Way They Do* (Free Press, 1998).

115   *By the time you have reached:* There is a vast literature on personality. In recent years this has been particularly resurgent, with fresh evidence of a single core structure and lifelong stability based upon genetic foundations. An authoritative overview of all aspects of the field can be found in Robert Hogan et al., eds., *Handbook of Personality Psychology* (Academic Press, 1997).

115   *In the jargon of marketing:* The USP idea comes from applying the logic of frequency-dependent selection to personality variation.

115   *Chimps, incidentally:* See J. E. King and A. J. Figuerodo, "The Five Factor Model Plus Dominance in Chimpanzee Personality" (*Journal of Research in Personality,* vol. 31, 1997, pp. 257–271).

116   *This is nature's way:* The pervasive influence of serotonin on a variety of social phenomena is discussed by Oliver James in *Britain on the Couch* (Century, 1997).

116   *It is the drug:* The delayed departure of many a CEO is a theme in Jeffery Sonnenfeld, *The Hero's Farewell* (Oxford University Press, 1988).

118   *Lars is from Sweden:* In Lars's and Annette's case history material are composite invented cases, drawn from various real cases from counseling

individuals with the aid of data from the NEO-PI-R, one of the leading Big Five personality measures. The Big Five are the key dimensions of human personality: neuroticism, extroversion, openness, agreeableness, and conscientiousness.

119   *In tribal societies:* Some of the First Nations tribes of America had different chieftains for war and peace. It could be said that the same thing happened in Britain, which found Winston Churchill to be the man of the hour during World War II but quickly dumped him in peacetime.

120   *Selection systems:* A complete exploration of the core idea of selection and its wide applicability can be found in Gary Cziko, *Beyond Miracles: Universal Selection Theory and the Second Darwinian Revolution* (MIT Press, 1995).

121   *Max dePree:* This business leader espoused his philosophy in a Stockton Lecture at the London Business School, 1993. See also his *Leadership Is an Art* (Dell, 1990).

121   *Many businesses have benefited:* The leadership succession at IBM and the benefits of internal choice in other firms are discussed in James Collins and Jerry Porras, *Built to Last* (Random House, 1994).

122   *Leaders are often narcissistic:* See Michael Maccoby, "Narcissistic Leaders: The Incredible Pros. The Inevitable Cons" (*Harvard Business Review*, January–February 2000, pp. 69–77).

123   *Leaders have two main:* "Task" and "people" dimensions of leadership functioning have dominated the literature for many decades. See "Leadership" and related entries in N. Nicholson, ed., *Encyclopedic Dictionary of Organizational Behavior* (Blackwell, 1995). There is a parallel with Rob Goffee and Gareth Jones's analysis of corporate culture in terms of the two dimensions of solidarity and sociability in *The Character of a Corporation* (HarperCollins, 1998).

5. PLAYING THE RATIONALITY GAME

128   *In the real world:* See Stuart Sutherland, *Irrationality: The Enemy Within* (Constable, 1992). For an economist's view of the problem, see Robert Frank, *Luxury Fever: Why Money Fails to Satisfy in an Era of Excess* (Free Press, 1999); see also Milton Leontiades, *Myth Management* (Blackwell, 1989).

128   *These ideas were less:* The indivisible relationship between human nature, magical belief, and the natural world is analyzed in Ariel Glucklich, *The End of Magic* (Oxford University Press, 1997). See also Robert Hinde, *Why Gods Persist: A Scientific Approach to Religion* (Routledge, 1999).

129   *It supports our belief:* Empirical research on illusory beliefs in free will is discussed in Denise C. Park, "Acts of Will" (*American Psychologist*, vol. 54, 1999, p. 461) and in other articles on this theme in the same journal issue. A nondualistic conception of the human place in nature is provided by Edward O. Wilson, *Consilience: The Unity of Knowledge* (Knopf, 1998). The

possibility that dualism is part of our hardwiring is called the "theory theory" by ethnopsychologists, who study basic human psychology in cross-cultural settings. See Angeline Lillard, "Ethnopsychologies: Cultural Variations in Theories of Mind" (*Psychological Bulletin*, vol. 123, 1998, pp. 3–32).

129　*The Miraculous Organ:* See Paul M. Checkland, *The Engine of Reason, the Seat of the Soul* (MIT Press, 1995).

130　*We tend to talk:* At the time of writing, computing based on DNA is being developed. This will no doubt increase computing power, but it may not simulate the brain any more than a series of electronic switches would.

131　*In addition, it has one:* The father of evolutionary psychology, Robert Trivers, has recently been exploring the universality of self-deception in human decision processes; see *The Logic of Self-Deception* (Weidenfeld and Nicolson, 2000).

132　*The Fall of Barings Bank:* Detailed accounts are to be found in Stephen Fay, *The Collapse of Barings* (Arrow Books, 1996), and John Gapper and Nicholas Denton, *All That Glitters* (Hamish Hamilton, 1996). For a more dramatized and less balanced account, see Nick Leeson's autobiographical version in *Rogue Trader* (Little, Brown, 1996)—also the title of a movie.

137　*The Feeling Factor:* See Robert Langs, *The Evolution of the Emotion-Processing Mind* (International Universities Press, 1996). See also Ross Buck, "The Biological Affect," *Psychological Review*, vol. 106. pp. 301–336 and Daniel Goleman, *Emotional Intelligence* (Bantam Books, 1995).

137　*Once our emotions are:* For a general psychological analysis of emotion, see Keith Oatley, *Best Laid Plans* (Cambridge University Press, 1992). Organizational perspectives are given in Stephen Fineman, ed., *Emotion in Organizations* (Sage, 1993). Darwin's work on emotional expression is annotated by a modern expert on the subject, Paul Ekman, in *The Expression of the Emotions in Man and Animals,* 3d ed. (HarperCollins, 1998). See also A. R. Damasio, *The Feeling of What Happens* (Heinemann, 2000).

140　*It has been said:* This view of autism has been discussed from an evolutionary perspective by Simon Baron Cohen in "Engineering and Autism: Is There a Link?" (Darwin seminar series at the London School of Economics, July 1997). See also his book *Mindblindness: An Essay on Autism and Theory of Mind* (MIT Press, 1995).

145　*Creative people:* See Michael A. West, *Developing Creativity in Organizations* (British Psychological Society, 1997), and Margaret A. Boden, *The Creative Mind* (Basic Books, 1990).

146　*The alternatives, too:* See Paul Carroll, *Big Blues: The Unmaking of IBM* (Crown, 1993), and Robert Slater, *Saving Big Blue* (McGraw-Hill, 1999).

147　*Risk Taking to Avoid Loss:* Loss aversion as a key to understanding risk behavior is discussed in Zur Shapira, *Risk Taking: A Managerial Perspective* (Russell Sage Foundation, 1995).

148   *The people prepared:* Animals' redoubling investments has been called the "Concorde" effect after the famous cost overrun on the Anglo-French supersonic airliner. See also Ulrich Müller-Herold, "Risk Management Strategies: Before Hominization and After" (*Journal of Risk Research,* vol. 3, 2000, pp. 19–30).

150   *Feedback, Failure:* See David Cannon, "Cause or Control? The Temporal Dimension in Failure Sense-Making," *Journal of Applied Behavioral Science,* vol. 35, pp. 416–438. Dietrich Dörner, *The Logic of Failure* (Perseus Books, 1996), and Joyce Fortune and Geoff Peters, *Learning from Failure: The Systems Approach* (Wiley, 1995).

152   *Failure avoidance:* Chris Argyris calls defensive routines "single-loop learning" that requires "double-loop learning" if we are to break out of them. In a conversation (1998) he told me that he considered single-loop learning to be a hardwired human tendency. See his *Knowledge in Action* (Jossey-Bass, 1993) and *Organizational Learning,* 2d ed. (Blackwell, 1994).

154   *It is estimated:* See Paul C. Nutt, "Surprising but True: Half the Decisions in Organizations Fail" (*Academy of Management Executive,* vol. 13, 1999, pp. 75–90). Psychological evidence for confidence illusions first figured in the work of Ellen J. Langer; see her *The Psychology of Control* (Sage, 1983).

154   *Psychological experiments:* Work by myself and colleagues at the London Business School has demonstrated that traders scoring high on control illusion in a computer game are rated as worse performers and make less money than their colleagues. See Mark Fenton-O'Creevy et al., "Trading on Illusions" (working paper, London Business School, 2000). "Behavioral finance," which focuses on the limits to economic rationality and predictive models based upon it, is booming as a new academic field. See Richard H. Thaler, *The Winner's Curse: Paradoxes and Anomalies of Economic Life* (Russell Sage Foundation, 1992).

154   *Media tycoon:* On Robert Maxwell, see Manfred F. R. Kets de Vries, *Life and Death in the Executive Fast Lane* (Jossey-Bass, 1995).

155   *Success fuels confidence:* The failure of success is increasingly being studied; see Pino G. Audia et al., "The Paradox of Success" (*Academy of Management Journal,* vol. 43, 2000, in press).

155   *Research shows that:* See M. H. Kernis et al., "Persistence Following Failure: The Interactive Role of Self-awareness and the Attributional Basis for Negative Expectancies" (*Journal of Personality and Social Psychology,* vol. 43, 1982, pp. 1184–1191).

156   *Tricks of the Mind:* An extensive inventory of human errors and illusions is to be found in Massimo Piatelli-Palmarini, *Inevitable Illusions: How Mistakes of Reason Rule Our Minds* (Wiley, 1994). See also John S. Hammond et al., "The Hidden Traps in Decision Making" (*Harvard Business Review,* September–October 1998, 47–52).

156 *The brain does contain:* See Brian Butterworth, *The Mathematical Brain* (Macmillan, 1999).

157 *We do have superb:* See R. Byrne, *The Thinking Ape: The Evolutionary Origins of Intelligence* (Oxford University Press, 1995), Alan Baddeley, *The Psychology of Memory* (Basic Books, 1976), and N. L. Stein et al., eds., *Memory for Everyday and Emotional Events* (Erlbaum, 1995).

158 *We love lists:* Our evolved addiction to lists is related to a range of other simplifying cognitive heuristics; see Nick Chater, "The Search for Simplicity: A Fundamental Cognitive Principle?" (*Quarterly Journal of Experimental Psychology,* vol. 52, 1999, pp. 273–302).

160 *Sampling and Statistical Errors:* See Livia Markóczy and Jeff Goldberg, "Women and Taxis and Dangerous Judgements: Content Sensitive Use of Base-Rate Information" (*Journal of Managerial Economics,* vol. 19, 1998, pp. 481–494).

161 *How is it possible:* See Paul Slovic, "The Construction of Preference" (*American Psychologist,* vol. 50, 1995, pp. 364–371).

163 *This is a well-known:* See A. V. Banerjee, "A Simple Model of Herd Behavior" (*Quarterly Journal of Economics,* vol. 107, 1992, pp. 797–817).

163 *In one species of grouse:* See Matt Ridley, *The Red Queen* (Viking, 1993).

164 *The Irrational Gift:* See Sutherland, *Irrationality: The Enemy Within* (Constable, 1992).

164 *We are great at guesswork:* See Guy Claxton, *Hare Brain, Tortoise Mind* (Fourth Estate, 1998). See also West, op cit., and Boden, op cit.

165 *Watch how you manage:* Michael Frese has pioneered the method of "error management" to overcome these difficulties; see "Error Management in Training" in S. Bagnara et al., eds., *Organizational Learning and Technical Change* (Springer-Verlag, 1996).

6. GENE POLITICS

168 *The fiendish:* See Dean Takaheshi and Dean Starkman, " 'Melissa' Dragnet Unsettles Virus Writers" (*The Wall Street Journal,* April 6, 1999).

169 *Romantic involvement:* One exception in discussing sexuality at work is Stephen Ackroyd and Paul Thompson, *Organizational Misbehavior* (Sage, 1999). See also the articles on romance at work by Sharon Foley and Gary N. Powell, and Gwen Jones in *Journal of Organizational Behavior* (vol. 20, no. 7, 1999).

169 *Sexual attraction to nonkin:* Walking on two legs was a great evolutionary innovation. It left hands free for carrying materials over distances, throwing projectiles, making tools, being artistic, and many other abilities, but it had this rather awkward design implication; see Colin Tudge, *The Day Before Yesterday* (Jonathan Cape, 1995).

169 *Falling in love:* Romantic love lasts for around a two-year period—long

enough to put the relationship through its paces and establish fertility. This is induced by a socially conditioned release of dopamine phenylethylamine and oxytocin, according to Professor Cindy Hazan of Cornell University, who conducted five thousand interviews and medical tests in thirty-seven cultures (reported by John Harlow, "True Love Is All Over in 30 Months," in the *Sunday Times* [London], July 25, 1999). See also Geoffrey Miller, *The Mating Mind* (Heinemann, 2000).

170   *In hunter-gatherer societies:* See Roger M. Keesing and Felix M. Keesing, *New Perspectives in Cultural Anthropology* (Holt, Rinehart & Winston, 1971).

171   *The event is commonplace:* There is also a dark side to these gene politics: evolutionists have noted the strong association of stepparenting and child physical abuse. See Margot Wilson and Martin Daly, *The Truth About Cinderella* (Weidenfeld & Nicolson, 1998).

172   *IBM started out:* See Paul Carroll, *Big Blues: The Unmaking of IBM* (Crown, 1993).

172   *Family firms can also:* A range of issues in family succession is discussed in Kelin E. Gersick, John A. Davis, Marion McCollom Hampton, and Ivan Lansberg, *Generation to Generation: Life Cycles of Family Business* (Harvard Business School Press, 1997).

173   *This is especially a problem:* Characteristics of entrepreneurs are discussed in my article "Personality and Entrepreneurial Leadership," (*European Management Journal*, vol. 16. 1998, pp. 529–539). See also Edgar H. Schein, "Entrepreneurs: What They're Really Like" (*Vocational Education Journal*, vol. 64, 1994, pp. 42–44).

174   *Beyond Our Kin:* Various examples of "as if" business behavior are described in Geoffrey James, *Giant Killers* (Orion, 1997).

176   *What happened in British:* See the report of the M&S saga by Peggy Hollinger, "Greenburg to Retire Early at M&S," *The Financial Times*, November 27, 1998.

178   *Here we see:* The "parent-adult-child" frame for looking at interpersonal relations—"Transactional Analysis"—was a restatement of Freudian psychodynamics by Eric Berne in *Games People Play* (Grove Press, 1964) and subsequent books. Evolutionary psychology can be seen as giving a fresh biogenetic perspective to these ideas. See also Christopher Badcock, "PsychoDarwinism: The New Synthesis of Freud and Darwin," in C. Crawford and D. L. Krebs, eds., *Handbook of Evolutionary Psychology* (Erlbaum, 1998).

179   *These rehearsals are essential:* The child development literature contains much research on the rehearsal role of play. For the evolutionary perspective, see Michael J. Boulton and Peter K. Smith, "The Social Nature of Play Fighting and Play Chasing," in Jerome H. Barkow et al., eds., *The Nurture Assumption* (Free Press, 1998). See also Judith Rich Harris for a more general discussion of the formative power of peer relationships at this stage.

179   *On one side of:* See Jeffery Young, *Forbes Greatest Technology Stories* (Wiley, 1998), and Robert X. Cringely, *Accidental Empires* (Addison-Wesley, 1992). Also see Collins and Porras, *Built to Last* (Random House, 1994), for an account of several of these business histories.

182   *Primitive critiques of:* The misreading and blatant misuse of Darwin's ideas by politicians, social philosophers, and leaders are documented in some detail by Mike Hawkins in *Social Darwinism in European and American Thought, 1860–1945* (Cambridge University Press, 1997).

182   *Our genes may be selfish:* The idea of the survival of the group depending upon selfless behavior has long been a hotly debated issue. Strictly speaking, there is no group interest in the selfish gene, apart from behaviors that advance the interests of kin and clan. However, it is plain that this is a looped program, for genes do help themselves by helping the group. These arguments are thoroughly explored in Elliott Sober and David Sloan Wilson, *Unto Others: The Evolution of the Psychology of Unselfish Behavior* (Harvard University Press, 1998).

184   *We are certainly hardwired:* See Roy F. Baumeister and Mark R. Leary, "The Need to Belong: Desire for Interpersonal Attachments as a Fundamental Human Motivation" (*Psychological Bulletin,* vol. 117, 1995, pp. 497–529).

185   *This points to:* The idea of a biological predisposition to internalize conscience was contained in Freud's idea of the superego. Robert Frank develops an analysis of the payoffs that underlie morality in his book *Passions Within Reason* (Norton, 1988).

186   *Men who grew up:* Unpublished findings from my own research database on birth order and personality show that the nurturance (agreeableness) dimension of personality is the only one of the Big Five to show clear effects with family size. This is consistent with other research.

186   *The extended families:* The greater value placed on nurturance in African management is embodied in the notion of *ubuntu;* see Mfuniselwa John Bhegu, *Ubuntu: The Essence of Democracy* (Anthroposophic Press, 1998). A similar value profile has been seen as characteristic of Islamic management; see University of Bradford Management Centre, U.K., Proceedings from the Third Arab Management Conference, July 1995.

187   *Contract Sensitivity:* See L. Cosmides and J. Tooby, "Cognitive Adaptations and Social Exchange," in J. K. Barkow et. al., eds, *The Adapted Mind* (Oxford University Press, 1992), on human sensitivity to contract violations.

188   *In hunter-gatherer communities:* See Carleton S. Coon, *The Hunting Peoples* (Cape, 1972).

188   *Self-deception gives:* See R. Trivers, *The Logic of Self-Deception* (Wiedenfeld and Nicolson, 2000), and Gary Johns, "A Multi-Level Theory of Self-Serving Behavior in and by Organizations" (*Research in Organizational Behavior,* vol. 21, 1999, pp. 1–38). An extension of the logic of self-deception is the "just world" hypothesis—the well-known psychological phenomenon of

people inclining to see people's misfortunes as being deserved in some way or other; see M. J. Lerner and D. T. Miller, "Just World Research and the Attribution Process" (*Psychological Bulletin,* vol. 85, 1959, pp. 1030–1051).

190   *Nowhere is this instinct:* Primo Levi, *If This Is a Man* (Abacus, 1979). The fundamentals of trading from an evolutionary perspective are also explored at length in Matt Ridley, *The Origins of Virtue* (Viking, 1996).

190   *This is well known:* See James Dow and Gary Gorton, "Noise Trading, Delegated Portfolio Management and Economic Welfare" (*Journal of Political Economy,* vol. 105, 1997, pp. 1024–1050).

192   *Evolutionary biologists:* See Karl Sigmund, *Games of Life: Explorations in Ecology, Evolution and Behaviour* (Oxford University Press, 1993).

193   *In a series of computer-simulated:* Many writers have discussed the evolutionary significance of these experiments conducted by George Axelrod. See, for example, Ridley and Dawkins, *The Selfish Gene* (Oxford University Press, 1976).

193   *Psychologists have done experiments:* See Frank, op cit., for a review of this work. Frank's own work shows that even minimal acquaintance greatly increases the likelihood of cooperation in prisoner's dilemma games; see his "The Evolution of One-Shot Cooperation" (*Ethology and Sociobiology,* vol. 14, 1993, pp. 247–256).

194   *Naturally friendly creatures:* Konrad Lorenz controversially drew parallels between humans and other species in *On Aggression* (Methuen, 1966). Comparative ethology and observations of primates seem to support some of Lorenz's assertions. See, for example, Robert A. Hinde, *Biological Bases of Human Social Behaviour* (McGraw-Hill, 1974), and Frans de Waal, *Chimpanzee Politics* (Jonathan Cape, 1982) and *Peacemaking Among Primates* (Harvard University Press, 1989).

194   *Human aggression is aroused:* Early writers on sociobiology gave a misplaced emphasis to territoriality and aggression—for example, Robert Ardrey in *The Territorial Imperative* (Collins, 1966). Hunter-gatherer peoples inhabit a home range within which they have no concept of personal space.

197   *In this case the macho:* This police force reformed its promotion practices after our investigation.

197   *From what dark well:* See Richard Wrangham and Dale Peterson, *Demonic Males: Apes and the Origins of Human Violence* (Houghton Mifflin, 1996), Barbara Ehrenreich, *Blood Rites: Origins and History of the Passions of War* (Virago, 1997), and Joanne Bourke, *An Intimate History of Killing* (Granta, 1999).

197   *Sports give license:* Business accounts of these feelings can be found in stories of takeover battles, such as the RJR Nabisco saga in Bryan Burrough and John Helyar, *Barbarians at the Gate* (Arrow Books, 1990), and the Fairfax

struggle in Glenn Burge and Colleen Ryan, *Corporate Cannibals* (Heinemann, 1992). See also Al Ries and Jack Trout, *Marketing Warfare* (McGraw-Hill, 1990), in which the authors talk about defensive, offensive, flanking, and guerrilla warfare. Military metaphors also figure in union-management conflicts, as described in my book with John Kelly and Jean Hartley, *Steel Strike* (Batsford, 1983).

197 *We (again, mainly men):* Psychologist Geoffrey Miller has collected a large body of evidence on the almost limitless range of criteria on which males will enact competitive achievement. See *The Mating Mind* (Heinemann, 2000).

200 *Intertribal conflict:* Raiding is a highly evolved behavior in humans and other primates. See Richard Wrangham, "Is Military Incompetence Adaptive?" (*Evolution and Human Behavior,* vol. 20, 1999, pp. 3–17).

201 *The alternative is to organize:* Something of this spirit is described in Manfred Kets de Vries' description of teamwork among the pygmy people in "High-Performance Teams: Lessons from the Pygmies" (*Organizational Dynamics,* Winter 1999, pp. 66–77).

203 *They don't actually* do *anything:* See Jon R. Katzenbach and Douglas K. Smith, "The Wisdom of Teams" (*Harvard Business Review,* March–April 1993, pp. 110–120).

204 *Like musicians in a jazz orchestra:* There has recently been much interest in jazz as a metaphor for organizing. See special issue on "Jazz Improvisation and Organizing" (*Organization Science,* vol. 9, no. 5, 1998).

204 *We cannot expect every:* Michael West has analyzed the deficiencies of work groups in business and their need for "reflexivity," i.e., self-examination. See M. A. West, "Reflexivity and Work Group Effectiveness" in M. A. West, ed., *Handbook of Work Group Psychology* (Wiley, 1996, pp. 555–579).

7. THE GOSSIP FACTORY

208 *In normal human society:* A child of six has memorized around 13,000 words, rising to 60,000 by adulthood. For a complete account of the evolution of language, see Steven Pinker, *The Language Instinct* (Morrow, 1994).

208 *Analysis of managers':* Managers' oral communication preferences were first highlighted in classic observational study by Henry Mintzberg, described in his *The Nature of Managerial Work* (Harper and Row, 1973). See also Colin P. Hales, "What Do Managers Do? A Critical Review of the Evidence" (*Journal of Management Studies,* vol. 23, 1986, pp. 88–115).

210 *It has been argued:* The use of stories in business is described in Thomas A. Stewart, "The Cunning Plots of Leadership" (*Fortune,* September 7, 1998, pp. 78–79). See also Ian McGregor and John G. Holmes, "How Storytelling Shapes Memory and Impressions of Relationship Events over Time" (*Journal of Personality and Social Psychology,* vol. 76, 1999, 403–419).

210 *Stories are memorable:* See also J. W. Kelly, "Storytelling in High-Tech

Organizations: A Medium for Sharing Culture" (*Journal of Applied Communication Research*, vol. 13, 1985, pp. 45–58).

213 *Alan C. Dershowitz:* He spoke in these terms on a BBC Radio 4 program, "Tell Me a Story," February 14, 1999.

213 *Psychologists call it:* See L. Ross, "The Intuitive Psychologist and His Shortcomings: Distortions in the Attribution Process," in L. Berkovitz, ed., *Advances in Experimental Social Psychology*, vol. 10 (Academic Press, 1977).

215 *This is everyday mind reading:* See Andrew Whiten, ed., *Natural Theories of Mind: Evolution, Development and Simulation of Everyday Mindreading* (Blackwell, 1991).

217 *How do we do this?:* The universal and involuntary nature of emotion-revealing expressions of face and body, and how people try to conceal and dissemble with them, is described by Paul Ekman in *Telling Lies* (Norton, 1985 & 1994).

217 *Unfortunately, when it comes:* See Neil R. Anderson, "Eight Decades of Employment Interview Research: A Retrospective Meta-Review and Prospective Commentary" (*European Work and Organizational Psychologist*, vol. 2, 1992, pp. 1–32).

218 *People unconsciously mirror:* See J. Bremmer and H. Rodenberg, *A Cultural History of Gesture* (Cornell University Press, 1991).

219 *Speaking and Listening:* See E. M. Eisenberg and H. L. Goodall, *Organizational Communication* (St. Martin's Press, 1993).

221 *Face-to-face communication:* See John Chadwick-Jones, *Developing a Social Psychology of Monkeys and Apes* (Psychology Press, 1998).

221 *The problem starts because:* Problems with e-mail communication were presented to the Academy of Management in Boston in August 1997 by Catherine Durnell in "Information Problems in Dispersed Teams." See also Lee Sproull and S. Kiesler, "Reducing Social Context Cues: The Case of Electronic Mail" (*Management Science*, vol. 32, 1986 pp. 1492–1512).

223 *Gossip as Grooming:* See Robin Dunbar, *Gossip, Grooming and the Evolution of Language* (Faber and Faber, 1996), and W. G. Runciman et al., eds., *Evolution of Social Behaviour Patterns in Primates and Man* (Oxford University Press, 1996).

226 *This is a typically male agenda:* See Deborah Tannen, *You Just Don't Understand* (Morrow, 1990), and *Talking from Nine to Five* (Avon Books, 1995). See also the research by Kathryn Waddington of the University of the South Bank, London, reported in *Times Higher Education Supplement*, July 10, 1998.

226 *The Rule of 150:* See Andrew Whiten and Richard W. Byrne, eds., *Machiavellian Intelligence* (Clarendon Press, 1988) and *Machiavellian Intelligence II* (Cambridge University Press, 1997).

227   *This idea led psychologist:* See Dunbar, op cit.

229   *The Politics of the Network:* See Lee Sproull and Sarah Kiesler, *Connections: New Ways of Working in the Networked Organization* (MIT Press, 1991). Mathematicians are discovering that a small number of connections between local networks creates enormous linkage possibilities across huge populations; see Duncan Watts, *Small Worlds: The Dynamics of Networks Between Order and Randomness* (Princeton University Press, 1999).

232   *Assassination:* See Patty Sotirin and Heidi Gottfried, "The Ambivalent Dynamics of Secretarial 'Bitching': Control, Resistance and Identity" (*Proceedings of the Academy of Management,* 1997, pp. 448–452).

234   *On a wider stage:* The social functions of award ceremonies and top-ten-type listings is analyzed in the work of N. Anand; see his "When Market Information Constitutes Fields" (*Organization Science,* in press, 2000).

235   *Rumor Is Improvised News:* See K. Davis, "Managerial Communication and the Grapevine" (*Harvard Business Review,* March 1953, pp. 43–49).

238   *Consider the contrast:* See the report on the TWA/Swissair comparison in Richard Tomkins, "Moments That Build or Destroy Reputations" (*Financial Times,* 17 September 1998). See also Norman R. Augustine, "Managing the Crisis You Tried to Prevent" (*Harvard Business Review,* November–December 1995, pp. 147–158).

239   *Fear of loss of control:* See S. Rosen and A. Tesser, "On the Reluctance to Communicate Undesirable Information: The MUM Effect" (*Sociometry,* vol. 33, 1970, pp. 253–263).

240   *Why and How Organizations:* Exploration of various of these themes can be found in chapters of F. M. Jablin et al., eds., *Handbook of Organizational Communication* (Sage, 1987).

243   *But contact between top management:* Personal communications from Raymond Ackerman and his management team, April 1999.

8. LIFE IN CAMP

246   *Monkeys are truly social:* See J. Chadwick-Jones, *Developing a Social Psychology of Monkeys and Apes* (Psychology Press, 1998).

246   *Chimpanzees have culture:* See W. C. McGrew, *Chimpanzee Material Culture: Implications for Human Evolution* (Cambridge University Press, 1992).

246   *The way in which we organize:* One of the first books to analyze the adaptive challenge in evolutionary terms was Kenneth W. Boulding, *Ecodynamics: A New Theory of Societal Evolution* (Sage, 1978). The new complexity theorists of the Santa Fe Institute have also explored adaptive "fitness" dynamics; see, for example, Stuart Kauffman, *At Home in the Universe: The Search for the Laws of Complexity* (Oxford University Press, 1995).

248   *The demands of a business:* See Wayne F. Cascio, "Downsizing: What Do We Know? What Have We Learned? (*Academy of Management Executive,* vol. 7, 1993 pp. 95–104).

250   *Size: How Big:* See T. J. Allen, "Size of Workforce, Morale and Absenteeism" (*British Journal of Industrial Relations,* vol. 20, 1982, pp. 83–100).

252   *Among business organizations:* On IBM, see R. Slater, *Saving Big Blue* (McGraw Hill, 1999) and P. Carroll, *Big Blues* (Crown, 1993).

252   *Other companies have tried:* A number of articles in *Academy of Management Review,* vol. 25, no. 1, 2000, explore the issue of corporate identity.

254   *"Organic" businesses:* The organic-mechanistic distinction between types of firms was first noted in a classic study by Tom Burns and G. M. Stalker, *The Management of Innovation* (Tavistock, 1966). The chimp colony study is reported in Margaret Powers, *The Egalitarians—Human and Chimpanzee* (Cambridge University Press, 1991), and the organizational interpretation of it in Barbara Pierce and Rod White, "The Evolution of Social Structure: Why Biology Matters" (*Academy of Management Review,* vol. 24, 1999, pp. 843–853).

254   *Percy Barnevik:* Barnevik discussed his business inspiration in a seminar at the London Business School on April 23, 1997.

255   *Hierarchy: Flexing the Muscles:* One of the first to note the universality of hierarchy in human societies through history was G. P. Murdock in "The Common Denominator of Cultures" (1945; reprinted in G. P. Murdock, ed., *Culture and Society,* University of Pittsburgh Press, 1965). On the failures of idealistic community, see S. Pinker, How the Mind Works (Norton, 1997).

255   *The urge to create:* I have explored the implications of EP for career systems in two articles; see Nigel Nicholson, "Career Systems in Crisis: Change and Opportunity in the Information Age" (*Academy of Management Executive,* vol. 10, 1996, pp. 40–51) and "Motivation-Selection-Connection: An Evolutionary Model of Career Development" in Maury Peiperl et al., eds., *Career Frontiers: New Concepts of Working Life* (Oxford University Press, 1999, pp. 54–75).

257   *Matrix Management:* See S. M. Davis and P. R. Lawrence, "Problems of Matrix Organization" (*Harvard Business Review,* May–June 1978, pp. 131–142), and C. A. Bartlett and S. Ghoshal, "Matrix Management: Not a Structure, a State of Mind" (*Harvard Business Review,* May–June 1990, pp. 138–145).

258   *New Corporate Forms:* See G. D. Dees et al., "The New Corporate Architecture" (*Academy of Management Executive,* vol. 9, 1995, pp. 7–18).

259   *The so-called boundaryless:* See John A. Byrne, "Jack: A Close-Up Look at How America's #1 Manager Runs GE" (*Business Week,* June 8, 1998, pp. 41–51).

259   *"Hollow" and "modular":* See S. Tully, "The Modular Corporation" (*Fortune,* February 3, 1993, pp. 106–108).

259   *The term* virtual organization: See Henry W. Chesbrough and David J.

Teece, "When Is Virtual Virtuous? Organizing for Innovation" (*Harvard Business Review*, January–February 1996, pp. 65–73), and N. Anand, "Sound Organization: Radical Organisation Design the Nashville Way" (working paper, London Business School, 1999).

260 *Telecommuting, for example:* See Yehuda Baruch and Nigel Nicholson, "Home, Sweet Work" (*Journal of General Management,* vol. 23 (2), 1997, pp. 15–30).

260 *There is a health warning here:* Robert Kraut et al., "Internet Paradox" (*American Psychologist,* vol. 53, 1998, pp. 1017–1031).

263 *In many contemporary:* See B. W. Schneider, "The People Make the Place" (*Personnel Psychology,* vol. 4, pp. 437–453) and M. Jordan et al., "Testing Schneider's ASA Theory" (*Applied Psychology: An International Review,* vol. 40, 1991, pp. 47–53).

265 *How is this possible?:* On selectionism, see G. Cziko, *Beyond Miracles* (MIT Press, 1995).

267 *Almost without exception:* Common themes have been noted by Jeffrey Pfeffer in *The Human Equation: Building Profits by Putting People First* (Harvard Business School Press, 1998).

269 *Cultures are defined by:* See Thomas E. Teal and A. A. Kennedy, *Corporate Cultures: The Rites and Rituals of Corporate Life* (Addison-Wesley, 1982).

272 *Magic and ritual are not:* See A. Glucklich, *The End of Magic* (Oxford University Press, 1997).

272 *The cave art and:* See B. Cooke and F. Turner, eds., *Biopoetics: Evolutionary Explorations in the Arts* (Icus Books, 1999). See also Paul G. Bahn, *Cambridge Illustrated History of Prehistoric Art* (Cambridge University Press, 1998).

274 *Can we already see:* See Gary Hamel, "Bringing Silicon Valley Inside" (*Harvard Business Review,* September–October, 1999, pp. 70–84). See also P. Bronson, *The Nudist on the Late Shift* (Secker and Warburg, 1999).

276 *We are surrounded by:* See, for example, William Bridges, *Jobshift: How to Prosper in a Workplace Without Jobs* (Addison-Wesley, 1994).

276 *All this reflects an increasing:* The vision of the new career has been discussed by Charles Handy in his writings; see, for example, *The Age of Unreason* (Hutchinson, 1981) and *The Empty Raincoat* (Hutchinson, 1984).

278 *The history of postagrarian:* See Jonathon Kingdon, *Self-Made Man and His Undoing* (Simon & Schuster, 1993).

Index